Springer Japan KK

T. Inoue · W.D. Pennie (Eds.)

Toxicogenomics

With 60 Figures, Including 7 in Color

Springer

Tohru Inoue, M.D., Ph.D.
Director
Center for Biological Safety and Research
National Institute of Health Sciences
1-18-1 Kamiyoga, Setagaya-ku,
Tokyo 158-8501, Japan

William D. Pennie, Ph.D.
Director
Molecular and Investigative Toxicology
Drug Safety Evaluation, Pfizer PGRD
Groton, Connecticut
USA

Cover: Toxicogenomics, by eliciting the counterbalancing gene expressions behind homeostasis, defines the mechanism of toxicology, thereby opening the road to predictive toxicology.

ISBN 978-4-431-67001-8 ISBN 978-4-431-66999-9 (eBook)
DOI 10.1007/978-4-431-66999-9

Library of Congress Cataloging-in-Publication data applied for.

Printed on acid-free paper

Typesetting: Authors

SPIN: 10867200

Foreword

The meteoric rate at which the human genome is being sequenced has presented to the research community a vast array of newly discovered genes, which in turn has catalyzed an even more dramatic effort to decipher this voluminous data set into understanding how genes function both individually and in complex pathways that regulate the biochemistry of life. A compendium of gene expression data, enhanced by complete proteomic analysis, will enable investigators to probe into the complexities of the mechanisms of normal genetic and metabolic pathways and, subsequently, how disease occurs when they malfunction. The new science of toxicogenomics combines genomic, proteomic, and informatics technologies, and biological research can now foresee a time when there will be a full comprehension of the complex dynamic mechanisms of genetics, biochemistry, and physiology.

The inherent power of toxicogenomics derives from an amalgamation of multiple scientific disciplines that were originally drawn together to facilitate sequencing the three billion bases that comprise the human genome. Traditionally, the science of toxicology has been founded upon empirical codification of dose-related effects. It has evolved to studies that are directed towards understanding the mechanisms by which individual agents cause their effects in humans. Due to technical limitations, this process has been relatively slow, since it has accomplished one chemical or one effect at a time. Also, to best grasp a full understanding of the mode of action of a single chemical, it is also necessary to gather data from structurally and functionally related isomers and their time- and dose-dependent biological effects. These results must be collated to develop an assessment of the mechanism of toxic action for a given chemical. Confounding our understanding of the toxic effects of a single chemical are often a plethora of additional environmental factors and stressors such as ultraviolet and ionizing radiation, biological agents, and dietary and lifestyle components that can contribute to the development of disease and make toxicity assessments more problematic. Therefore, to better understand the full effects of a toxic chemical, the actions of all agents must be characterized to a progressively greater depth in order to properly fathom the biochemical and genetic complexity of the cells, tissues, or organism in which adverse effects are manifest. The great promise of toxicogenomics is that it enables the simultaneous analysis of thousands of genes and proteins in a single experiment and thereby can greatly facilitate the growth of mechanistic knowledge of normal biochemical pathways as well as determination of how they go awry and cause disease.

This is especially true of time- and dose-dependent microarray data that will be reflective of phases of chemical activity. Initial responses of organisms or tissues to chemical or drug exposure within 24–48 h, at doses that are not acutely toxic, may provide data on specific genes involved in the early pharmacologic action of a drug or chemical. As exposure to the stressor or agent is increased in time or dose, toxicity or cellular injury will become progressively obvious and various adaptive functions will be expressed. The use of microarrays and concomitant proteomic analysis should thus provide the opportunity to search for signature pathways of toxic injury. Such data will allow insight into the mode or mechanism of toxic injury and will also provide a means of distinguishing array patterns indicative of the adverse effect of the agent. If array data can be "phenotypically anchored" to conventional indices of toxicity (histopathology, clinical chemistry, etc.), it will be possible to search for evidence of injury prior to its clinical or pathological manifestation. This approach could lead to development of early biomarkers of toxic injury and thus may also help to resolve issues related to interspecies extrapolation and variation in susceptibility across individuals.

While mRNA analysis is a potentially powerful tool for recognizing chemical-induced effects, analysis of protein sequence, structure, and post-translational modification provides advantages more clearly reflecting the actual current state of activity of the cell or tissue. Promising new methods in proteomics are emerging, including the capacity to profile proteins with surface-enhanced laser desorption time-of-flight (SELDI-TOF) mass spectrometry and antibody arrays. Correlation between changes in mRNA and protein levels will offer insights into the function of genes, and will serve as a guide in the search for protein biomarkers of chemical exposure and predictive toxicity.

The promise of this new technology is such that it can be used to generate data on large numbers of chemicals and exposure conditions and can be used to develop an unprecedented knowledge base that can be used to guide future research, improve environmental health, and aid in regulatory decisions. Development of the knowledge base must proceed incrementally and requires a worldwide collective effort of many individuals, institutions, and countries.

Toxicogenomic knowledge may grow slowly in the early stages as various standards emerge. Developers and users must have the patience and dedication to stay the course, since incremental growth will at some point give way to an autocatalytic exponential growth and the field of toxicology will be profoundly changed. Given the vast numbers and diversity of drugs, chemicals and environmental stressors, the diversity of species in which they act, the time and dose factors that are critical to the induction of beneficial and adverse effects, and the diversity of phenotypic consequences of exposures, it is only through the development of a rich knowledge base and its availability to all of the scientific community that the fields of toxicology and environmental health can rapidly advance. Concomitant with development of the data/knowledge base must be the evolution of informatics (computational and statistical) and data mining tools (query algorithms, relational interfaces, etc.) and the individuals trained to apply them.

Toxicogenomics promises to have a revolutionary impact on environmental health, drug safety, and risk assessment. However, the scientific problems to be solved in these areas are longstanding and extremely complex, and therefore progress will take significant time and effort. First, toxicogenomics will contribute to understanding biological responses to environmental stressors and will identify agents that are a significant risk to human health. Toxicogenomics is a powerful tool for improving human risk assessment because it will measure specific changes in gene expression in humans and other species that are exposed to drugs or other agents. Careful data analysis could identify similar patterns in different species, leading to a "signature" for a given pathway of toxicity.

Toxicogenomics can improve assessment of exposure to toxicants since specific mRNA expression signatures will be the basis of new tools for human exposure assessment. Using these signatures, it may be possible to identify the agent and dose to which individuals or populations have been exposed. Eventually this approach will help develop protein biomarkers of exposure and effect. Protein biomarkers of effect may also be important for early detection of environmentally induced disease. Toxicogenomics can also be used to identify susceptibility factors that influence an individual's response to environmental agents. It is clear that individuals differ in their susceptibility to environmentally related disease. Toxicogenomic studies can define gene expression patterns in nonsusceptible and susceptible individuals who have suffered adverse responses to drugs or chemicals. This work may lead to highly targeted gene expression and protein arrays that can be used to predict which individuals are likely to be adversely affected by different agents. These studies will also help elucidate the mechanisms of susceptibility to different agents, which will lead to safer and more effective therapies or drugs.

Thus, toxicogenomics presents an unprecedented opportunity to have a dramatic impact on environmental health and chemical and drug safety to all the people of the world. This information will dramatically change our understanding of human disease risk and will provide new opportunities for nations and individuals to protect their health by understanding their relative risk of disease. The program of the Toxicogenomics International Forum 2001 exemplifies the excitement of this new generation of research. Research presentations at the forum show the increasing blend of new models with the latest high-throughput technologies, and application of them to the most current genomic and proteomic research questions. It is through such combined efforts that toxicogenomics will make significant contributions to understanding the most basic biochemistry and genetics of life. These next few years will be a most exciting period in the history of biological research, and we have great anticipation of the new discoveries just over the horizon that will revolutionize our understanding of the biochemistry of life.

Raymond W. Tennant, Ph.D.
Director, National Center for Toxicogenomics
National Institute of Environmental Health Sciences
Research Triangle Park, NC, USA

Foreword

Successful outcomes of the international cooperative research program on a complete elucidation of the structure of the human genome are thought to revolutionize various areas of life sciences, particularly medical and pharmaceutical sciences. The application of genomic technologies to toxicology (toxicogenomics) is expected to allow a greater understanding of the adverse effect of a chemical on human populations in terms of both elucidating individual susceptibility to an adverse effect and determining the mechanism by which the chemical can cause the adverse effect. Thus, toxicogenomics is now regarded as a basic scientific tool and discipline to be applied to improving methodologies for risk analysis of environmental chemicals as well as facilitating the research and development of new drugs.

In this context, the first Toxicogenomics International Forum 2001 was convened to provide an opportunity for the exchange of information among participants from multidisciplinary areas in regard to the methodology, application strategies, and future perspectives of toxicogenomics in the field of risk analysis for chemicals and drugs.

Yuzo Hayashi, M.D., Ph.D.
President, Advisory Committee
Toxicogenomics International Forum

Preface

As the pace of new drug, agrochemical, and chemical discovery increases, traditional toxicology research is challenged to deliver quality candidate safety information without becoming the rate-limiting step in compound advancement. While discovery scientists focus attention on single biomolecules (or pathways) as targets, the scientist engaged in safety assessment has to attempt to predict or determine a novel compound's effect on a very wide range of biological endpoints. The comparatively recent application of microarray technology to toxicology (popularly referred to as toxicogenomics) is an example of where the "genome revolution" is being explored by toxicologists in the drive to keep pace with the discovery of new chemical entities.

The careful application of transcript-profiling technologies has the potential to assist research associated with safety assessment:

- An enhanced ability to extrapolate accurately between experimental animals and humans in the context of risk assessment.
- A more detailed appreciation of molecular mechanisms of toxicity.
- Facilitating faster screens for compound toxicity.
- The provision of new research leads.

The breadth of contributions to this volume is a testament to the multidisciplinary nature of the field of toxicogenomics. There are a variety of technologies used to profile the changes in expression of many thousands of genes simultaneously. These may consist of commercially available technical platforms or in-house constructed "broad-coverage" or hypothesis-based gene arrays. Contributions to this volume describe the practicalities of design, construction and utilization of these platforms. Microarrays find usage in a broad range of applications from characterizing a specific mechanism of toxicity (such as endocrine disruption) to attempting to predict the potential toxicological properties of novel chemicals based on gene expression "fingerprints." Such pattern recognition could facilitate the discovery and subsequent validation, of biomarkers useful for application in higher throughput approaches to toxicology testing.

The application of toxicogenomics is not without challenge, however. Technical issues around sensitivity, reproducibility and gene annotation need to be addressed to ensure we obtain the best possible data. The large volume data sets produced by this approach require innovative approaches in database design and querying especially if experiments performed on different technical platforms are to be compared. In addition, the statistical tools employed to analyze microarray results continue to evolve rapidly, and the plethora of available approaches can make comparison of analyzed data sets from different laboratories challenging. All of these issues will undoubtedly be resolved as researchers share their perspectives and experience; the contributions in this volume describe how individual laboratories are working singularly and in collaboration to utilize toxicogenomics to maximal advantage today. As our understanding of the utility of this approach matures, the appropriate usage of toxicogenomics for risk assessment purposes will become apparent.

The editors tried to cover the content as comprehensive as possible and to figure out the state-of-arts of this newly developing area, "toxicogenomics." Contributors of each book chapter are all active pioneers in the microarray technology, and being either invited at the international forum held in Tokyo on October 31 – November 1, 2001, or additionally requested to participate in the publication. We hope this would be a sort of milestone of this area.

Tohru Inoue, M.D., Ph.D.
Director
Center for Biological Safety
and Research
National Institute of Health Sciences
Setagaya-ku, Tokyo

William D. Pennie, Ph.D.
Director
Molecular and Investigative Toxicology
Drug Safety Evaluation, Pfizer PGRD
Groton, Connecticut

Contents

Introductory Keynotes

Strategy for Toxicogenomics

Technical Development (Genomics and Proteomics)

Experimental Applications (1) — Oncology Research

Experimental Applications (2) — Xenobiotic Metabolism

Experimental Applications (3) — Endocrine Disruptor Research

Future Prospects and Risk Assessment

Contributors

Aardema, Marilyn J., Miami Valley Laboratories, The Procter & Gamble Co., P.O. Box 538707, Cincinnati, OH 45252, USA. aardema.mj@pg.com

Adachi, Tetsuya, Core Research for Evolutional Science and Technology (CREST), Japan Science and Technology Corporation, Kawaguchi 332-0012, Japan

Anderson, Larry, Battelle Pacific Northwest National Laboratory, 902 Battelle Boulevard, Richland, Washington 99352, USA

Arai, Ken-ichi, Institute of Medical Science, University of Tokyo (IMSUT), 4-6-1 Shirokanedai, Minato-ku, Tokyo 108-8639, Japan. ken-ichi@ims.u-tokyo.ac.jp

Arizono, Koji, Faculty of Environmental and Symbiotic Sciences, Prefectural University of Kumamoto, 3-1-100 Tsukide, Kumamoto-shi 862-8502, Japan, and CREST, JST. kjarzn@pu-kumamoto.ac.jp

Bartosiewicz, Matthew, Department of Bioengineering, University of California, San Diego, 8656 B Via Mallorca Drive, La Jolla, CA 92037, USA. embartosiewicz@hotmail.com

Blumberg, Bruce, Department of Developmental and Cell Biology, 5205 BioScience II, University of California, Irvine, CA 92697-2300, USA. blumberg@uci.edu

Bostian, Keith A., Iconix Pharmaceuticals, 850 Maude Avenue, Mountain View, CA 94043, USA. kbostian@iconixpharm.com

Bradfield, Christopher A., McArdle Laboratory for Cancer Research, University of Wisconsin Medical School, Madison, Wisconsin 53706, USA. bradfield@oncology.wisc.edu

Buckpitt, Alan, Department of Molecular Biosciences, School of Veterinary Medicine, University of California, Davis, CA 95616, USA. arbuckpitt@ucdavis.edu

Cheng, Rita, Hematology/Oncology Section, Department of Pediatrics, Texas Children's Cancer Center, M3-3320, Baylor College of Medicine, Houston, TX 77030, USA

Clothier, Bruce, MRC Toxicology Unit, Hodgkin Building, University of Leicester, Lancaster Road, Leicester LE1 9HN, UK

Davies, Reginald, MRC Toxicology Unit, Hodgkin Building, University of Leicester, Lancaster Road, Leicester LE1 9HN, UK

Dix, David J., Reproductive Toxicology Div. (MD-72), US Environmental Protection Agency (EPA), National Health and Environmental Effects Research Laboratory, Research Triangle Park, NC 27711, USA. dix.david@epa.gov

Edmunds, John W., Syngenta Central Toxicology Laboratory, Macclesfield, Cheshire, SK10 4TJ, UK

Engelberg, Alan, Iconix Pharmaceuticals, 850 Maude Avenue, Mountain View, CA 94043, USA

Furness, Mike, Iconix Pharmaceuticals, 850 Maude Avenue, Mountain View, CA 94043, USA

Gant, Timothy W., MRC Toxicology Unit, Hodgkin Building, University of Leicester, Lancaster Road, Leicester LE1 9HN, UK. twg1@le.ac.uk

Goodman, Jay I., Department of Pharmacology and Toxicology, Michigan State University, B-440 Life Sciences Building, East Lansing, MI 48824, USA. goodman3@pilot.msu.edu

Grewal, Anoop, Silicon Genetics, 2601 Spring Street, Redwood City, CA 94063, USA. anoop@sigenetics.com

Handa, Hiroshi, Frontier Collaborative Research Center, Tokyo Institute of Techology, 4259 Nagatsuta, Yokohama 226-8503, Japan

Hayes, Kevin R., McArdle Laboratory for Cancer Research, University of Wisconsin Medical School, Madison, Wisconsin 53706, USA

Hirose, Masao, Division of Pathology, National Institute of Health Sciences, 1-18-1 Kamiyoga, Setagaya-ku, Tokyo 158-8501, Japan. m-hirose@nihs.go.jp

Ichiishi, Eiichiro, Laboratory of Genome and Protein Functions, First Department of Medicine, Kyoto Prefectural University of Medicine, 465 Kajii-cho, Kawara-machi-dori Hirokouji, Kamigyo-ku, Kyoto-shi 602-5266, Japan. ichiishi@koto.kpu-m.ac.jp

Iguchi, Taisen, Center for Integrative Bioscience, Okazaki National Research Institutes, CREST and JST, 38 Nishigonaka, Myodaiji, Okazaki-shi 444-8585, Japan. taisen@nibb.ac.jp

Inoue, Tohru, Center for Biological Safety and Research, National Institute of Health Sciences, 1-18-1 Kamiyohga, Setagaya-ku, Tokyo 158-8501, Japan. tohru@nihs.go.jp

Ishikawa, Toshihisa, Department of Biomolecular Engineering, Graduate School of Bioscience and Biotechnology, Tokyo Institute of Technology, 4259 Nagatsuta-cho, Midori-ku, Yokohama-shi 226-8501, Japan. tishikaw@bio.titech.ac.jp

Jarnagin, Kurt, Iconix Pharmacauticals, 850 Maude Avenue, Mountain View, CA 94043, USA. kjarnagin@iconixpharm.com

Judah, David J., MRC Toxicology Unit, Hodgkin Building, University of Leicester, Lancaster Road, Leicester LE1 9HN, UK

Kai, Toshinori, Faculty of Environmental and Symbiotic Sciences, Prefectural University of Kumamoto, 3-1-100 Tsukide, Kumamoto-shi 862-8502, Japan

Kanno, Jun, Division of Cellular and Molecular Toxicology, Center for Biological Safety and Research, National Institute of Health Sciences, 1-18-1 Kamiyoga, Setagaya-ku, Tokyo 158-8501, Japan. kanno@nihs.go.jp

Kohara, Yuji, Center for Genetic Resource Information, National Institute of Genetics, Mishima, Japan. ykohara@lab.nig.ac.jp

Komiyama, Masatoshi, Department of Bioenvironmental Medicine, Graduate School of Medicine, Chiba University, 1-8-1 Inohana, Chiba-shi 260-8670, Japan. komiyama@med.m.chiba-u.ac.jp

Kondo, Akihiro, Biotechnology Research Laboratory, Takara Shuzo Co., Ltd., 3-4-1 Seta, Ohtsu, Shiga 520-2193, Japan. kondoa@takara.co.jp

Lipson, Doron, MRC Toxicology Unit, Hodgkin Building, University of Leicester, Lancaster Road, Leicester LE1 9HN, UK

Lomas, Lee O., Ciphergen Biosystems Inc., 6611 Dumbarton Circle, Fremont, CA 94555, USA. lomas@ciphergen.com

MacGregor, James T., US Food and Drug Administration, National Center for Toxicological Research, 5600 Fishers Lane, Rockville, MD 20857, USA. macgregorj@cder.fda.gov

Masutomi, Naoya, Division of Pathology, National Institute of Health Sciences, 1-18-1 Kamiyoga, Setagaya-ku, Tokyo 158-8501, Japan

Mendrick, Donna L., Vice President, Toxicology Program, Gene Logic Inc., 708 Quince Orchard Road, Gaithersburg, MD 20878, USA.
dmendrick@genelogic.com

Mizutani, Takeshi, Center for Integrative Bioscience, Okazaki National Research Institutes, CREST and JST, 38 Nishigonaka, Myodaiji, Okazaki-shi 444-8585, Japan

Moggs, Jonathan G., Syngenta Central Toxicology Laboratory, Macclesfield, Cheshire, SK10 4TJ, UK.

Morgan, D. Gwyn, 230 Windermere Ave; Wayne, PA 19087, USA

Mori, Chisato, Department of Bioenvironmental Medicine, Graduate School of Medicine, Chiba University, 1-8-1 Inohana, Chiba-shi 260-8670, Japan.
mori@med.m.chiba-u.ac.jp

Morris, James, Battelle Pacific Northwest National Laboratory, 902 Battelle Boulevard, Richland, Washington 99352, USA

Murphy, Tracy C., Syngenta Central Toxicology Laboratory, Macclesfield, Cheshire, SK10 4TJ, UK

Natsoulis, George, Iconix Pharmaceuticals, 850 Maude Avenue, Mountain View, CA 94043, USA

Orphanides, George, Syngenta Central Toxicology Laboratory, Macclesfield, Cheshire SK10 4TJ, UK. george.orphanides@syngenta.com

Pearson, Cecelia, Iconix Pharmaceuticals, 850 Maude Avenue, Mountain View, CA 94043, USA

Penn, Sharron G., Aeomica, Sunnyvale, California 94085, USA

Pennie, William D., Syngenta, Central Toxicology Laboratory, Alderley Park, Macclesfield, Cheshire, SK10 4TJ, UK.; (present address) Molecular and Investigative Toxicology, Drug Safety Evaluation, Pfizer PGRD, Groton, Connecticut, USA. william_d_pennie@groton.pfizer.com

Pettit, Syril D., ILSI Health and Environmental Sciences Institute, 1 Thomas Circle, NW, Ninth Floor, Washington, D.C., 20005-5802, USA. spettit@ilsi.org

Pounds, Joel, Battelle Pacific Northwest National Laboratory, 902 Battelle Boulevard, Richland, Washington 99352, USA. joel.pounds@pnl.gov

Rank, David R., Aeomica, Sunnyvale, California 94085, USA

Riley, Joan, MRC Toxicology Unit, Hodgkin Building, University of Leicester, Lancaster Road, Leicester LE1 9HN, UK

Robinson, Denise E., ILSI Health and Environmental Sciences Institute, 1 Thomas Circle, NW, Ninth Floor, Washington, D.C., 20005-5802, USA; (present address) Global Technology Development, Pfizer Global Research and Development, 50 Pequot Avenue, MS6025-C4269, New London, CT 06320 USA. Denise_Robinson-Gravatt@groton.pfizer.com

Robinson, Susan, MRC Toxicology Unit, Hodgkin Building, University of Leicester, Lancaster Road, Leicester LE1 9HN, UK

Sasser, Lyle, Battelle Pacific Northwest National Laboratory, 902 Battelle Boulevard, Richland, Washington 99352, USA

Shibutani, Makoto, Division of Pathology, National Institute of Health Sciences, 1-18-1 Kamiyoga, Setagaya-ku, Tokyo 158-8501, Japan. shibutan@nihs.go.jp

Smith, Andrew G., MRC Toxicology Unit, Hodgkin Building, University of Leicester, Lancaster Road, Leicester LE1 9HN, UK. ags5@le.ac.uk

Styles, Jerry A., MRC Toxicology Unit, Hodgkin Building, University of Leicester, Lancaster Road, Leicester LE1 9HN, UK

Suzuki, Atsuko, Center for Integrative Bioscience, Okazaki National Research Institutes, CREST and JST, 38 Nishigonaka, Myodaiji, Okazaki-shi 444-8585, Japan

Tabb, Michelle M., Department of Developmental and Cell Biology, 5205 BioScience II, University of California, Irvine, CA 92697-2300, USA

Takagi, Hironori, Division of Pathology, National Institute of Health Sciences, 1-18-1 Kamiyoga, Setagaya-ku, Tokyo 158-8501, Japan

Takeda, Ken, Faculty of Pharmaceutical Sciences, Tokyo University of Science, Ichigaya, Shinjuku-ku, Tokyo 162-8601, Japan. takeda@ps.kagu.sut.ac.jp

Thomas, Russell S., Aeomica, Sunnyvale, California 94085, USA. rthomas@kalypsys.com

Tominaga, Nobuaki, Ariake National College of Technology, Omuta 836-8585, Japan

Tran, Karen, Aeomica, Sunnyvale, California 94085, USA

Turton, Nicola J., MRC Toxicology Unit, Hodgkin Building, University of Leicester, Lancaster Road, Leicester LE1 9HN, UK

Uneyama, Chikako, Division of Pathology, National Institute of Health Sciences, 1-18-1 Kamiyoga, Setagaya-ku, Tokyo 158-8501, Japan

Ura, Kazuhiro, Faculty of Environmental and Symbiotic Sciences, Prefectural University of Kumamoto, 3-1-100 Tsukide, Kumamoto-shi 862-8502, Japan and CREST, JST. kazu@host1.mhs-unet.ocn.ne.jp

Watanabe, Hajime, Center for Integrative Bioscience, Okazaki National Research Institutes, CREST and JST, 38 Nishigonaka, Myodaiji, Okazaki-shi 444-8585, Japan. watanabe@nibb.ac.jp

Wong, Kwong-Kwok, Hematology/Oncology Section, Department of Pediatrics, Texas Children's Cancer Center, M3-3320, Baylor College of Medicine, Houston, TX 77030, USA. kkwong@txccc.org

Yoshikawa, Megumi, Department of Biomolecular Engineering, Graduate School of Bioscience and Biotechnology, Tokyo Institute of Technology, 4259 Nagatsuta-cho, Midori-ku, Yokohama-shi 226-8501, Japan

Yoshikawa, Toshikazu, First Department of Medicine, Kyoto Prefectural University of Medicine, 465 Kajii-cho, Kawara-machi-dori Hirokouji, Kamigyo-ku, Kyoto-shi 602-5266, Japan

Zastrow, Gina M, McArdle Laboratory for Cancer Research, University of Wisconsin Medical School, Madison, Wisconsin 53706, USA

Zhou, Changcheng, Department of Developmental and Cell Biology, 5205 BioScience II, University of California, Irvine, CA 92697-2300, USA

Acknowledgments

As a result of the terrorist attacks on September 11, 2001, Joseph J. DeGeorge, J. Christopher Corton, and several other contributors could not participate in this publication, but the editors appreciate their encouragement. Editorial reviews by Katsuhide Igarashi, NIHS, Makoto Shibutani, NIHS, and Hajime Watanabe, NIBB, are appreciated. Without the editorial assistance of Noriko Katsu, this publication would not have been possible.

Introductory Keynotes

Introduction:

Toxicogenomics - a new paradigm of toxicology

Tohru Inoue

Center for Biological Safety & Research, National Institute of Health Sciences
1-18-1 Kamiyohga, Setagayaku, Tokyo 158-8501, Japan

Summary. Molecular biology has enabled the elucidation of biological subjects with bilateral strategies, namely, an inductive approach and a deductive approach. Along with the development of the mouse whole-genome sequencing project, it has enabled elucidation of the science bilateral interrelationships between the toxicological phenotypes related to particular toxicants and expression profiles of pertinent genes induced by exposure to toxicants. While a conventional inductive approach permits exploration of the toxicological mechanism by cloning genes and analyzing gene and protein expression during the course of chemical exposure, the newly developed deductive approach potentially permits the elucidation of the toxicological phenotype(s) through gene expression.
Microarray technology has dramatically changed the time course of drug discovery in new drug development. Potential therapeutics can be screened for thousands of endpoints indicative of efficacy and adverse toxicity at one time using the microarray technology. Simultaneously, the same technology can be used to explore unique genomic "expression fingerprints", which can be used to group the biological effects of chemical actions at a various doses, time intervals, or target tissues, in a variety of animal species, into profiles as the bases of gene expression. Accumulation of the expression profiles (here and elsewhere) of whole genomes for reference chemicals for a variety of treatment conditions permits the establishment of an informatics profile (here and elsewhere) for reverse toxicology, which is conversely supposed to predict the toxicological phenotypes solely by analyzing gene expression. This translational introductory oversees the future prospects of how microarrays can be used in applied toxicology.

Key words. Toxicogenomics, DNA microarray, reverse science, reverse genetics, reverse toxicogenomics

DNA microarrays

As an introductory keynote to "Toxicogenomics", a discussion on what toxicogenomics can offer to conventional toxicology is given here in this paragraph. Toxicogenomics is based on DNA microarray and DNA chip technologies that are similar to those in other genome science fields (Lovett, 2000; Hamadeh et al., 2001; Storck et al., 2002) i.e., the DNA microarray fixed with cDNA by a DNA spotter, and hybridized with fluorescence-labeled cDNAs from tissue samples (Schena et al., 1995, 1996), and the DNA chip, on which a number of oligonucleotide probes are photolithographically synthesized, followed by hybridization of biotiny lated cDNAs from samples (Fodor et al., 1993). Originally, DNA microarray and DNA chip technologies have been used to analyze a large number of gene expressions, and thus, have been applied to such functional genomics fields as transcriptomics (Storck, et al., 2002) pharmacogenomics (Lloyed A, 2000), mutagenomics (Aardema and MacGregor, 2002), oncogenomics (Herrmann, et al., 2001), pathogenomics (Liefers, et al., 2001), and predictive diagnostic medicine based on clinical prognosis (Nakamura, 2001), and specifically, the latter DNA chip technology is a potentially powerful tool for identifying DNA sequences, thus, such inductive information has been applied widely in the research for single nucleotide polymorphism, SNP, in a variety of drug-metabolizing enzymes, etc., to establish an individualized "tailor-made pharmacology".

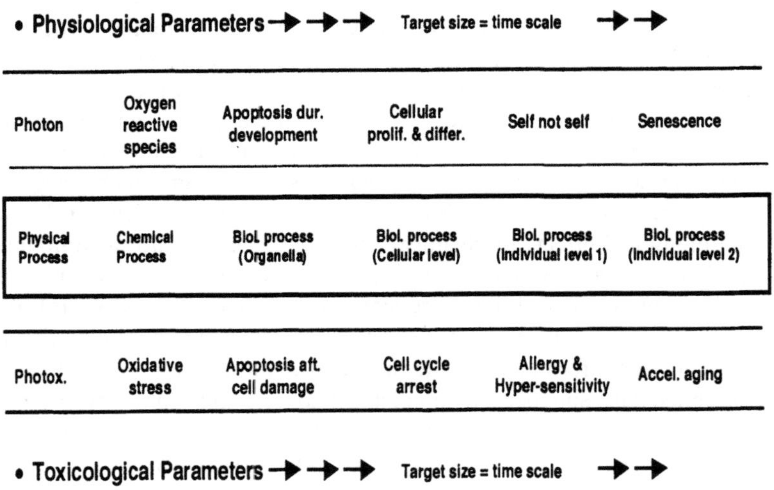

Fig. 1. Physiological parameters shown in the upper row vs. toxicological ones shown in the lower row along with the increase of participating target masses or time scales.

Microarray and /or DNA chip technologies applied in toxicology are called "toxicogenomics". Toxicogenomics can be applied bi-laterally, either inductively or deductively. Deductive approach of toxicogenomics shows a great, unexpected paradigm shift from conventional toxicology.

Toxicology and toxicogenomics

Before describing an over- view of toxicogenomics, what toxicology is, namely, its definition, entity, and scientific bases, should be reviewed first. Toxicology is an interdisciplinary area between biology/medicine and chemistry/physics. Key molecules participating in physiological responses and toxicological responses are presumably comparable (**Figure 1**), implying that physiological responses and toxicological responses may be a continuum. Pharmacology involves the identification of something available; on the other hand, toxicology involves the identification of not only the mechanism of toxicity but also clarifying a border of "nothing", i.e., NOEL, "no observed effect level", and/or NOAEL, "no observed adverse effect level". The goals of toxicology are to predict the effect of potential hazards on human health effects, and to identify the mechanism of toxicity, NOEL and/or NOAEL. In this regard, toxicogenomics is supposed to clarify comprehensively the border of "nothing". Although a prototype of "toxicogenomics" was developed in 1997 (Heller et al., 1997) to identify specific toxicological phenotypes, such as oxidative stress inducers, drug- metabolizing chemicals, and cell-cycle-specific modulators, comprehensive toxicogenomics became possible after the whole-genome sequencing project was accomplished in 2001. Because of the completion of the whole genome sequence, finally, the toxicology to predict "nothing" became possible.

Birth of reverse science & toxicology

In 1988, a new era of mouse genetics, reverse genetics, was started by generating the first knockout gene for mammalian species, murine int-2, by the group of Mario Capecchi's (Capecchi et al., 1988) and then Elizabeth Robertson's (Schwartzberg, et al, 1989). Thereafter, molecular biology has enabled the elucidation of biological subjects by bi-directional strategies, forward and reverse ones, i.e., the inductive and the deductive approach, respectively, where not only genes that possess a particular expression phenotype have been cloned by forward genetics, but also a number of genes of which functions were not known have been uncovered their function by reverse genetics, i.e., knockout technologies. The history of genetics teaches such bilaterally alternating strategies to strengthen scientific power. Thus, it is speculated that the inductive toxicology and deductive toxicology may complement each other.

Along with the development of the mouse whole-genome sequencing project, such bi-directional strategies for analysis became possible also in toxicology; the toxicologic phenotypes of particular toxicants and the expression profiles of pertinent genes reacting with the toxicants. While the inductive approach permits exploration of the toxicological mechanism by analyzing gene and protein expression during the course of toxicological testing, the deductive approach permits prediction of the toxicological phenotype(s) solely by analyzing the gene expression. Microarray and/or DNA chip technologies have enabled the survey of a large number of gene expressions after exposure to a toxicant. Both inductive and deductive approaches have enabled application of DNA chip and/or the microarray in toxicological analysis, i.e., "toxicogenomics". Toxicogenomics enables exploration of the toxicological mechanism by analyzing a large number of gene chips inductively, and opens a new era of reverse toxicology, which is supposed to predict possible toxicologic phenotypes by distinguishing the expression patterns of particular genes from accumulated expression profiles. The DNA chip and the microarray technologies for the identification of specific toxicity groups are commercially available already, e.g., metabolic enzyme inducers, growth factor & receptor-mediated transducers, xenobiotic ligands for nuclear receptors, stress-response-gene modifiers, and cell-cycle regulator modifiers.

Reverse toxicology

Similar to reverse genetics, reverse toxicology is supposed to identify toxicological phenotypes solely by examining their expression profiles. Such deductive use of microarray technology for toxicology is called "Reverse Toxicogenomics", where it is expected to predict toxicological phenotypes solely by analyzing whole gene expression (**Figure 2**). This technique is requires a

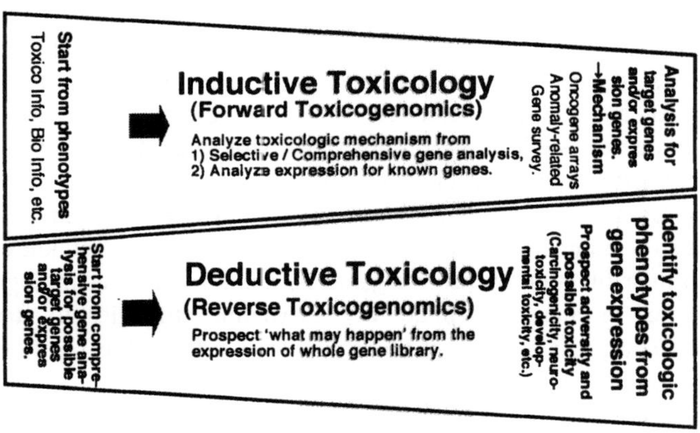

Fig 2. Structures of inductive toxicology vs. deductive toxicology. Former starts its analyses from toxicologic phenotypes toward the mechanism, whereas, the latter focuses in identigying toxicologic phenotypes solely from the gene expression profiling.

minimum number of animals, or even samples of in vitro-cultured cells after a relatively short period of exposure to a potential hazardous testing materials. The predictability by reverse toxicology depends upon the number of gene expression profiles accumulated, the number of phenotypes differentially linked to the gene expression profiles, and informatics linking such gene expressions and the phenotypes.

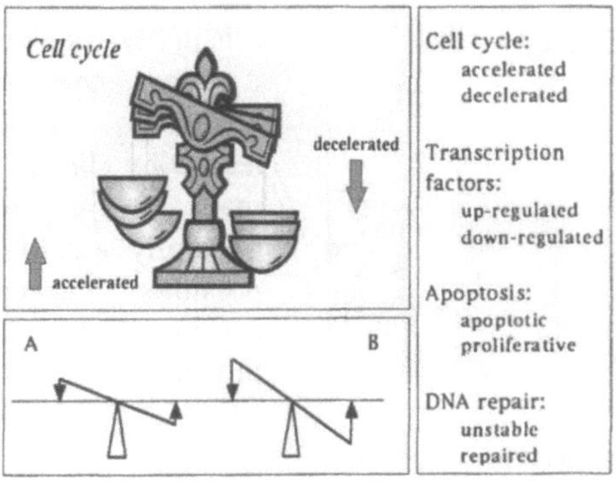

Fig. 3. Counter balancing gene expressions behind the homeostasis Visualization of different oscillatory balances A and B (left bottom).

At this moment, reverse toxicogenomics is still a theory. However, a variety of testing methods in toxicology will be replaced by reverse toxicogenomics, eventually. This strategy has the following advantages: it reduces the number of test animals, and the test period, and adopts simpler techniques by using established expression profile as new biomarkers rather than sophisticated methodologies requiring skill and experience. Furthermore, a specific proteome chip, expressing a series of specific genes, is supposed to function as "reverse proteomics" through a series of processes such as sample preparation, 2D gel electrophoresis, and mass-spectrometry for image analysis (Zhu et al., 2001). To set up an endpoint where NOEL or NOAEL exists, a traditional toxicology has been applied to incorporate something " invisible borders". Invisible borders are, in conventional toxicology, based on at least two major limitations: one in an endogenous factor(s) and the other in an exogenous factor. The former, for example, is hidden behind homeostasis, and the latter, for example, is behinbd a technological limitation. As shown in **Figure 3**, most living animals exhibit homeostasis between two (or more) counter- balancing vectors such as oxidation & reduction, apoptosis & anti-apoptosis, and acceleration & deceleration of cell cycle regulation. Since the counter-balancing counter-directional homeostasis, it appears static and one may not recognize the differences between one homeostatic stage, balanced at a low energy stage, to the other stage, balanced at a high energy stage (A and B in Figure 3). It is far more important to note that stage B is generally more risky. Toxicogenomics is expected to disclose such hidden homeostatic balances which are undetectable by conventional testing systems. The latter, an exogenous factor in a technological limitation, may be based on such resolution limit of light-microscopes, spectrophotometers, etc., and all

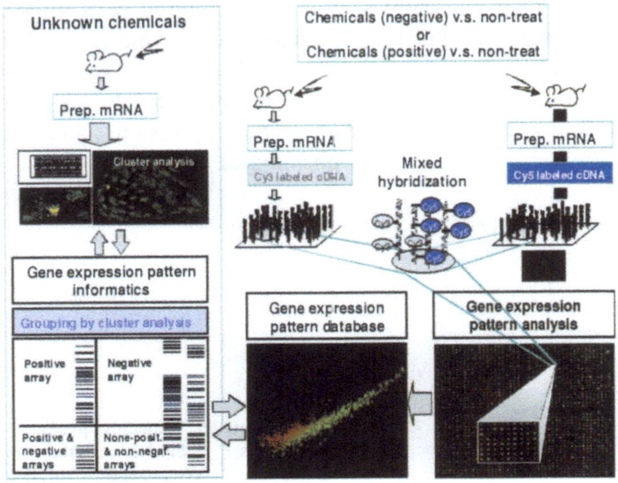

Fig. 4. Practical toxicity-predicting system based on the gene expression microarray.

technologies exhibit their resolution point, i.e., "invisible barrier". Toxicogenomics also has a limit in terms of technical sensitivity; however, it may overcome presently available resolution limits in many ways, and hopefully identify a possible specific toxicological profiling.

Practical approach

A sample of a chemical toxicity predicting system is shown in **Figure 4**. Mice are treated with, or without, a known toxic reference chemical (TRC), or treated with, or without, a known non-toxic reference chemical (NTRC). Then, the

Table 1: Possible toxicologic endpoints tested by in vitro or in vivo resources

Possibility of in vitro test	Toxicity Endpoint(s)
No [yes]*	*Morphogenesis, developmental anomaly*
No	*Identifying tissue-specific toxicity*
No/yes	*Epigenetic carcinogenesis (as modification of gene expression)*
	Metabolic activation
No	Hepatic activation leading to multi-tissue damage
Yes	Tissue-specific activation
	Receptor-mediated events
Yes/no	Neuronal tissue
Yes/no	Steroid hormonal tissue
Yes/no	Ah-receptors
Yes	*Cytotoxicity*
	Membrane activities
Yes	ion channels, ion pumps
	Inhibition of biochemical process
Yes	Uncoupling oxidative phosphorylation, inhibition of ATP production by redox cycling initiation to produce ROS
Yes	Oxidative phosphorylation to inhibit ATP production
Yes/no	Alteration of calcium homeostasis

*[yes] Limited possibility at this moment, e.g. whole embryo culture

messenger RNAs are extracted, and visualized with red color marker, cy3, for overexpression or with green color marker, cy5, for down-modulation. These color-labeled mRNAs will be processed into a competitive mixed-hybridization in a high-density hybridization array. Expression patterns are informatized in many ways. Along with accumulation of data to establish informatic profiles, specific gene clusters for TRC, NTRC, and those positive for both TRC + NTRC, andnegative for both TRC + NTRC, can be established. These databases can be compared with an expression profile that will be obtained from unknown chemicals (left box in Figure 4).

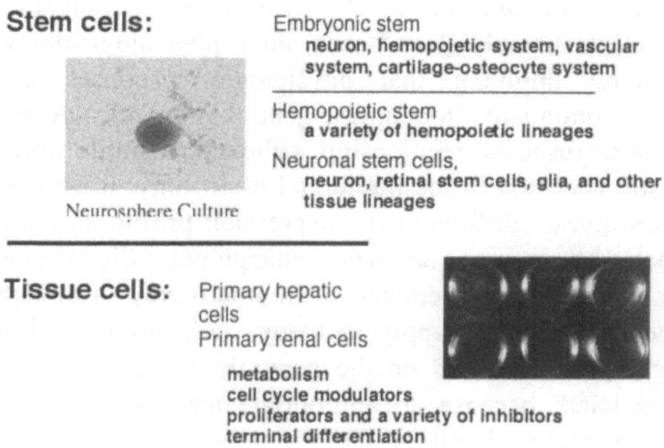

Fig. 5. A variety of in vitro resources for toxicologic gene expression array. Different cellular function for microarray analyses between stem cells and tissue cells. See text.

The technique requires only a limited number of animals, or even with cultured cells, in vitro, after relatively a short period of exposure. Depending upon the endpoints of toxicity aimed to focus on, even materials from in vitro culture may work efficiently (**Table 1**). As shown in Table 1, activities of membrane such as ion channels and ion pumps, the inhibitory effect of uncoupling oxidative phosphorylation, and the inhibition of ATP turnover by redox cycling, may be identified by the microarray. Possible toxicity related to developmental anomaly and morphogenesis (top of Table 1) may not be predictable by the use of in vitro cell culture; however, as seen in **Figure 5**, an in vitro system, for example, an embryonic stem (ES) cell, may predict some possible adverse effects of toxicity on the morphogenesis. Consequently, ES cells as well as hemopoietic and neuronal stem cells are particularly powerful tools for identifying the effect of toxicity on not only proliferation but also differentiation. Actually, one ES cell potentially corresponds to one individual; therefore, observing a microarray of ES cells may correspond to observing several millions of mice at the early developmental stage. Hepatic, renal and other types of primary cultured cells are limited but useful for observing such a variety of metabolic modulators, cell cycle regulators, and cell proliferation inhibitors and/or stimulators, in primary hepatic cells.

New paradigm of toxicology

Toxicogenomics, specifically reverse toxicogenomics, is about to open a new paradigm of toxicology from, at least, five aspects: first, the merging of such scientific borders as physiology and toxicology (also pharmacology and toxicology); second, a paradigm shift from "analog science" to "digital science"; third, visualization of hitherto unknown oscillatory changes behind homeostatic balances; fourth, comprehensive inter-species extrapolation; and lastly, a paradigm shift from inductive toxicology to deductive toxicology. As discussed previously in the first paragraph of this chapter, when we compare the aims of physiology, and toxicology, we find that they face opposite directions; however, participating molecules, seen in Figure 1, are presumably shared each other, physiologically as well as toxicologically, thereby implying that physiologic responses and toxicologic responses may be a continuum. In contrast, gene expression may not be continuum along with the dose response relationship, although a simple linear dose-response curve is generally accepted in the traditional toxicology. It appears to be clear that a different dose gives a different gene expression profile, in other words, the expression is expected to show not an analog change but a digital one, and a different dose behaves as a different chemical in the microarray. Although simple linear dose-response curves seem to apply in many cases, toxicological parameters may change discontinuously based on the genomic expression. This may be an advantage on one hand, because an appropriate array profiling of toxicologic responses can be eventually identified. On the other hand, there may still be a long way to go before reaching the final goal of defining the specific toxicologic array profiling for appropriate toxicologic phenotypes. The concept of safety borders, such as NOEL and NOAEL, may likely be re-established likely by means of such specific safety profiles. Visualization of invisible homeostatic balances is shown in the Figure 3. As was mentioned in the paragraph on oscillation strength, an additional new concept of "risk" may be re-established. Comprehensive 'interspecies'-extrapolation may be improved by informatics over different species such as mouse, rat, frog (Xenopus), and yeast. Interspecies extrapolation will be dealt with on a theoretical basis using a relatively small number of genes that are known to confer important allelic variations. This would result in better interspecies extrapolation, higher confidence of animal models, reduction in the number of animals needed for testing, shorter testing period, and most importantly, insights into pathways of toxicity and their mechanisms (US-EPA). Reverse toxicogenomics may be supported by the other four paradigm shifts mentioned above, and be used to predict toxicity and establish new concepts in risk assessment methodologies.

By means of toxicogenomics, we will be able to see a new toxicological world behind homeostasis and/or gene expression balance.

Acknowledgement: This review work was, in part, the product from the collaboration study with Yoko Hirabayashi, MD, PhD, Cellular & Molecular Toxicology Division, Center for Biological Safety and Research, NIHS, Japan.

References

1. Lovett RA: Toxicogenomics. Toxicologists brace for genomics revolution. Science 289: 536-537, 2000.
2. Hamadeh HK, Bushel P, Paules RS, Afshari CA.: Discovery in toxicology: mediation by gene expression microarray. Biochem Mol Toxicol 15: 231-242, 2001.
3. Storck T, von Brevern MC, Behrens CK, Scheel J, Bach A.: Transcriptomics in predictive toxicology. Curr Opin Drug Discov Devel 5: 90-97, 2002.
4. Schena M, Shalon D, Davis RW, Brown PO: Quantitative monitoring of gene expression patterns with a complementary DNA microarray. Science 270: 467-470, 1995 / Schena M, Shalon D, Heller R, Chai A, Brown PO, Davis RW: Parallel human genome analysis: microarray-based expression monitoring of 1000 genes. Proc Natl Acad Sci USA 93: 10614-10619, 1996.
5. Fodor SP, Rava RP, Huang XC, Pease AC, Homes CP, Adams CL: Multiplexed biochemical assays with biological chips. Nature 364: 555-556, 1993.
6. Lloyd A: Pharmacogenomics-Qualitative gene profiling using differential analysis of transcripts with alternative splicing profiles. Drug Discovery Today 5: 429-430, 2000.
7. Aardema MJ, MacGregor JT: Toxicology and genetic toxicology in the new era of "toxicogenomics": impact of "-omics" technologies. Mutat Res 499: 13-25, 2002.
8. Herrmann JL, Rastelli L, Burgess CE, Fernandez EE, Rothberg BE, Rothberg JM, Shimkets RA: Implications of oncogenomics for cancer research and clinical oncology. Cancer J 7: 40-51, 2001
9. Liefers GJ, Tollenaar RAEM, Nakamura Y, van de Velde CJH: Genetic cancer syndromes and large-scale gene expression analysis: applications in surgical oncology. Europ. J Surg Oncol 27: 343-348, 2001.
10. Nakamura Y: Isolation of disease-associated genes through genome analysis and their clinical application. Keio J Med 50: 134-140, 2001.
11. Kamataki T, Nunoya K, Sakai Y, Kushida H, Fujita K: Genetic polymorphism of CYP2A6 in relation to cancer. Mutat Res 428: 125-130, 1999.
12. Heller RA, Schena M, Chai A, Shalon D, Bedilion T, Gilmore J, Wooley DE, Davis RW: Discovery and analysis of inflammatory disease-related genes using cDNA microarrays. Proc Natl Acad Sci USA 94: 2150-2155, 1997.
13. Capecchi MR: Altering the genome by homologous recombination. Science 244: 1288-1292, 1989.
14. Schwartzberg PL, Goff SP, Robertson EJU: Germ-line transmission of a c-abl mutation produced by targeted gene disruption in ES cells. Science 246: 799-803, 1989.
15. Zhu H, Bilgin M, Bangham R, Hall D, Casamayor A, Bertone P, Lan N, Jansen R, Bidlingmaier S, Houfek T, Mitchell T, Miller P, Dean RA, Gerstein M, Snyder M.: Global analysis of protein activities using proteome chips. Science 293: 2101-2105, 2001
16. Zhu H, Snyder M: Protein arrays and microarrays. Curr. Opin. Chem. Biol. 5: 40-45, 2001.

Toxicogenomics in Genome Science World

Ken-ichi Arai

Institute of Medical Science, University of Tokyo (IMSUT), 4-6-1, Shirokanedai, Minato-ku, Tokyo 108-8639

Summary. About 500 genes have been listed as drug targets until recently. However, wealth of information provided by human genome analysis has dramatically changed the picture and we now have to deal with more than 30,000 genes. They are replicated, recombined, repaired, transcribed into RNA, spliced and translated into proteins in autonomous manner or in response to environmental stimuli. How to find the functions of these genes is the major task in functional genomics in post-sequencing era. Likewise, discovering disease related and/or therapeutic genes become the key issue to develop novel drugs and to predict their behavior and actions in human body. SNP analysis and expression profiling using microarray technology and proteomic analysis offer great promise from diagnostic aspect. Toxicogenomics employs similar tools available in genomic research. Progress of molecular and cellular biology and biotechnology in the last decades has altered the traditional concept of drugs based on organic chemistry. In addition to small compounds, proteins (biotech drugs), peptides, DNAs, RNAs (gene therapy) and cells (cell therapy) have been added to the drug catalogues. How to select and/or combine various therapeutic approaches optimized to particular diseases and to individuals is important for order-made (or tailored) medicine.

Key words. Molecular biology, paradigm shift, genome science

Lessons of Silicon Valley in Biotechnology

I would like to talk about briefly the trend in biomedical research (Table.1). The first stage was the establishment of molecular biology in early 1950s promoting fundamental discoveries which helped to establish the basic principle in DNA related research (Fig. 1). The second stage was opened by the invention of recombinant DNA technology which triggered the creation of biotechnology industry and bioventures (Fig. 2). We already talked about

Table 1

Paradigm Shift in Biomedical Research

From Molecular Biology to Genomic Medicine
Role of Translational Research
Deductive and Predictive Approaches
Toward Functional Genomics

NIH-based research and the FDA approval process in the east coast. In the 70s, the Silicon Valley of the west coast has been the prime place to develop biotechnology. In the late 70s to 80s, I had been working at Stanford and DNAX in the Bay area and I would like to point out that the the discovery and invention made at Stanford University teamed with entrepreneuership have been the driving force to create knowledge-based bioindustry. What was unique in Silicon Valley was that the discovery and the innovation were transformed effectively into biotechnology industry in a very concentrated area. Dr. Alejandro Zaffaroni, an inventor of ALZA's drug delivery system in 60s and the founder of DNAX, Affymax and Maxygen, once pointed out three critical components to promote bioventures: 1) An international university attracting scientists world wide. 2) Access to international airport within 30 minutes. 3) Favorable living conditions. The Silicon Valley has been funded by venture capitals as well as by big pharmaceutical companies. Again, the unique feature is that the venture capitalists and the big pharmas of the east coast worked together to create bioventures and/or bridge companies to develop effective biotech drugs.

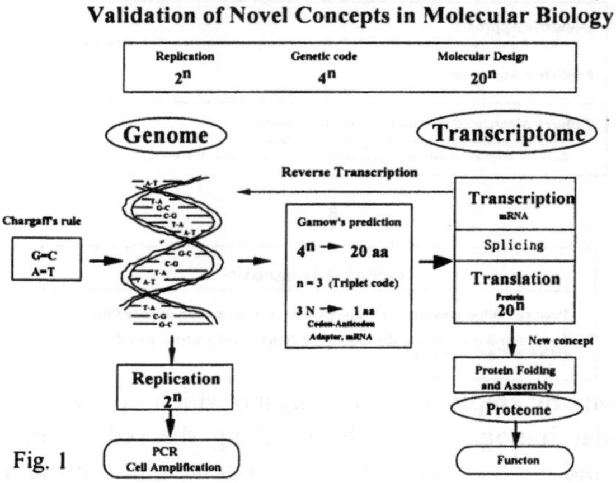

Fig. 1

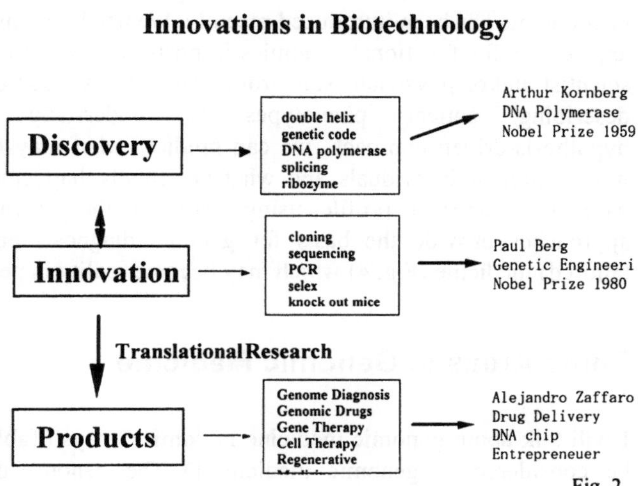

Fig. 2

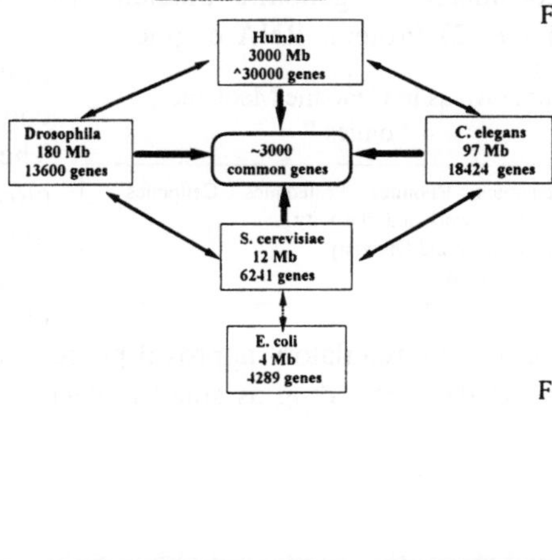

Fig. 3

Paradigm Shift in Biomedical Research

Then, another paradigm shift began in late 90s due to advances in genomic research. DNA is a micro-machine with nanometer scale and human genome consists of 3 x 10^9 base pairs with DNA scale of ~1 m. If we assume human body consists of ~10^{14} cells, this corresponds to ~100 grams of DNA with total length exceeding 10^{11} km. Monitoring the DNA of the entire body will take ~3 days even by employing the machine which operates by speed of light (Fig 3). Molecular biologists have been taking the reductionist approach studying the structure and the function of individual gene and/or protein in the past. However, whole-istic and integrative approach studying whole genome or proteome has been introduced recently. We need both individual and integrative approaches for functional genomics in post-sequence era. In deductive approach, scientist and/or physicians start from observations made on the bench or bedside describing patient phenotypes to understand pathophysiology. In hypothesis-driven approach, we can predict "what may happen" using genomic information of individuals and what is actually happening could be monitored based on expression profile, using DNA and the protein chips (Fig. 4). These approaches provide the basis for genome diagnosis enabling order-made (or tailored) medicine (Fig. 4) which may become reality in next decades.

Fig. 4

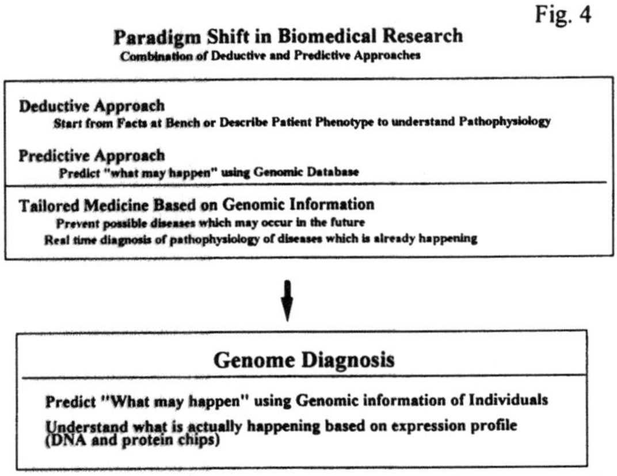

Three Areas in Genomic Medicine

I will talk about genomic medicine in "omics" days (Table 2). Three areas could be considered in genomic medicine:1) The genome diagnosis and genomic medicine. 2) Protein/ RNA engineering and proteomics for genomic drugs and molecular design etc. 3) Cell engineering for regenerative medicine. They are not necessarily separable and closely related each other. Based on these developments, the concept of drugs in the medical frontier changes dramatically which is an important agenda for the regulatory approval process in safety and efficacy considerations. From the days of a drug as small molecular weight compound based on organic

Table 2
Three Areas in Genomic Medicine
"omics"

Genomics , Rnomics , Proteomics , Cellomics
Signal Network and Plasticity
Universality and Diversity
System Biology

chemistry, recombinant proteins and antibodies were brought in as biotech drugs in the 70s. Now more than 50 drugs of a protein base have been approved and more recently, antibodies have been introduced to the market. Then comes the days of genomic drugs including gene therapy. This gene therapy is different from what we have been attempting for the last decade to replace altered gene by normal gene; rather it is an effort to achieve therapeutic effect using genes as a drug. Then there are still other drugs linked to cell therapy.

Genome Diagnosis, Functional Genomics and Protein-Protein Interaction

In the post-sequencing era, we can analyze at the DNA, RNA, and protein levels to understand gene function. Either we deal with individual genes by employing knock out or knock in, ribozyme, antisense, antibody, peptide, aptamer etc. The classical concept that a single gene causes genetic disease such an inborn error of metabolism or molecular disease, is still important. The interest more recently is the analysis of multiple genes that cause common diseases such as hypertension, cancer, and diabetes mellitus etc by using SNPs as polymorphic markers. This is a possibility that still may happen. In Japan, SNP discovery project directed

Fig. 5a **Protein Interactions in Signaling Cascade**

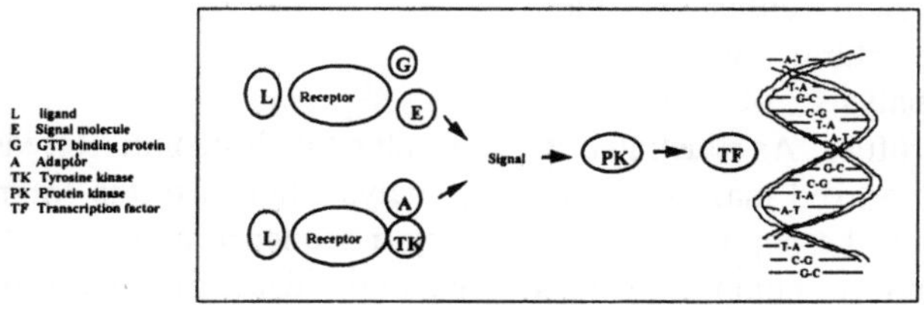

Reaction Type	Chemistry	Drug Development
Intermediary Metabolism	Organic Chemistry	organic compound agonist/antagonist
Template Reaction	Molecular Genetics	base-pair oligonucleotide
Signal Transduction	Molecular Recognition	protein interaction peptide

Allosteric Protein Network

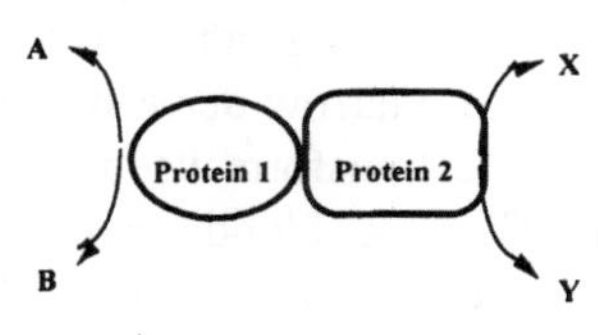

Molecular Recognition
Protein Interaction
Allostery
A B chemically related
XY chemically related
AB and XY Chmically unrelated
chaos or or order
Plasticity of the Network
Memory

Fig. 5b

Table 3 **Signal Network**

Multigene Family
Ligands
Receptors
Signal Cascade
Cross Talks
Cytoplasm-Nuclear Interaction
Target molecule

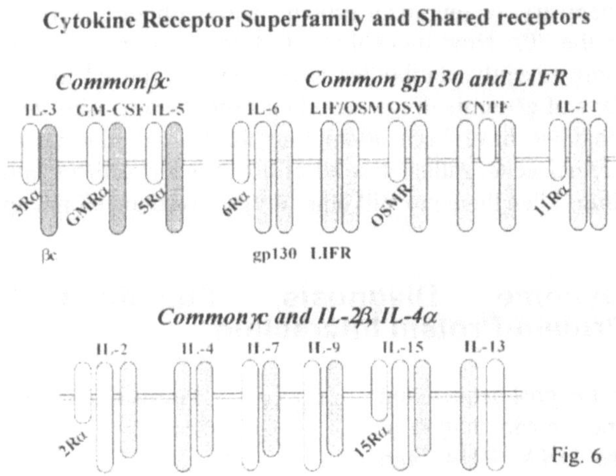

Fig. 6

by Dr. Yusuke Nakamura, the collaborative project of IMSUT · Human Genome Center, JST, and RIKEN Polymorphism Research Center, has already identified about 200,000 SNPs from the Japanese population. The next important activity is a proteome system either at the structural level and analysis of protein profiles. As illustrated in the Center for Proteomic Research, the major tool is an improved mass spectroscopic analysis linked to the fragmentation of the protein into the peptides. By obtaining a sequence of six or seven amino acids stretch and to compare with the genomic database, the protein can be identified. Therefore, we can know which part of the genome has been translated into proteins. Proteomic analysis would provide novel tools to characterize protein-protein interactions in cell signaling network (Figs. 5 and 6) and revolutionize the level of genomic medicine.

Human Development, Stem Cells and Cell Therapy

The third area is the use of cells as a drug. As illustrated in a textbook of gynecology, the cell fate in the formation of human body is determined within the first 7 weeks of human development. We can derive stem cells of various developmental potential including embryonic stem cells which provide future promise for regenerative medicine. Somatic cell therapy is an alternative approach such as by employing cord blood- and/or bone marrow-derived stem cells. Recently, the ability of mesenchymal stem cells to transform itself into various cell lineages have been recognized. At IMSUT, cord blood bank have been supported by donation department (Asahi Chemicals and Nipro) and has been operated for seven years to do research of cord blood stem cells. Each Cell Archive freezer can accommodate about 3,500 cord blood samples and if 20,000 cord blood samples become available, that is sufficient to cover the need of the Japanese population. In fact, this type of facility has been employed to support the life of the patient of nuclear accident that happened two years ago in Japan. The major challenge in cell therapy is to develop the technology to amplify stem cells without inducing cell differentiation (Cell PCR) or to induce cell differentiation into particular lineage. My laboratory, in collaboration with Dr. Hisao Masai of

Tokyo Metropolitan Institute of Medical Science, is developing Cell PCR technology by manipulating the basic machinery regulating DNA replication and cell cycle such as to trigger G1-S transition artificially, thereby resulting in amplification of stem cells.

Challenges in Drug Development

I would like to discuss the challenges we face in drug development: high risk (about 2% of lead compounds succeeded), high cost (about $100 million for clinical trial), and long time (about 13 years for approval). The cost was over $500 million for the private sector, and the public sector would have paid another $500 million. This process must be improved substantially. I would like to discuss what shall we do in Japan (Fig. 7)? The challenge we face in genomic medicine in the new environment is to discover how it differs from traditional drug development.

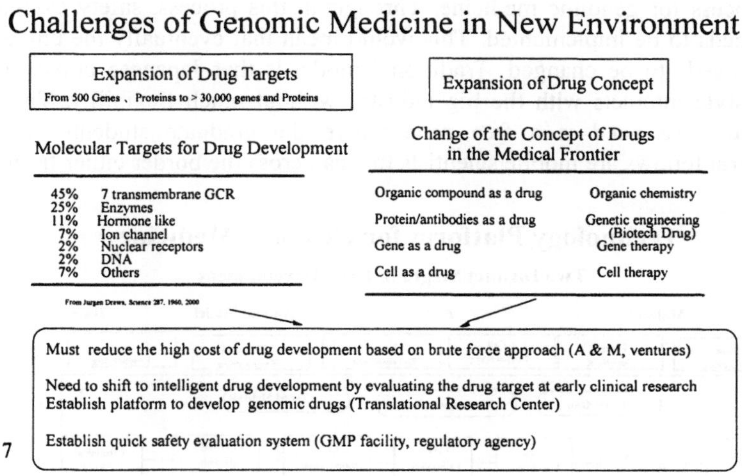

Fig. 7

According to Dr. Jurgen Drews, we have about 500 genes as drug targets thus far but still it costs a huge amount of money. Now human genome project disclosed that more than 30,000 genes are available. It is clear we cannot pursue drug development according to the old model, because it will suck up huge resources. It is not clear yet whether we should deal with these genes as a drug, as organic chemistry, or protein or genes or cells. We simply have to quickly evaluate their promise at early phase of clinical research by shifting to intelligent drug development. Now let me summarize a couple of points. Knowledge about human genome should be used to reduce rather than to increase the high cost of drug development that is based on the brute force approach. Many American or European companies are doing A & M in collaboration with the venture capitals.

This process does not work well in Japan and we need to establish common platform of genomic medicine (i.e. translational research and its centers) to develop genomic drugs. Also, we need to establish quickly the safety evaluation system in drug development. We need more established GMP-level facilities closely linked to the research community. We also need smarter process for safety evaluation, which means that a Japanese version of the FDA must be established. Fig. 8 shows a technology platform for genomic medicine including animal disease model and early clinical research. This whole process of drug development will take 10 years or more. How shall we develop this new industry? Two points I would like to make are, first, university-based research must be linked to the industry via bioventures. The first stage to create ventures corresponds to "proof of principle" stage which requires extensive translational research. The second one is the stage to develop company-based research and clinical trials. However, in Japan either in academia, industry and government, I found that these distinct stages which face problems unique to each are all mixed up. We have to build both the university-based and the company-based researches, if we are to develop this entire process for genomic medicine. Throughout this process, safety evaluation system needs to be implemented. This would mean that eventually the university structure needs to be changed. Traditional model is that Japanese university of German style interacts with the big industry without much flexibility. The new environment we need is a free zone where the graduate students and the postdoctoral fellows are mobile scientists that can cross the border either from

Fig. 8 **Technology Platform for Genomic Medicine**

Two Distinct Stages in Drug Development

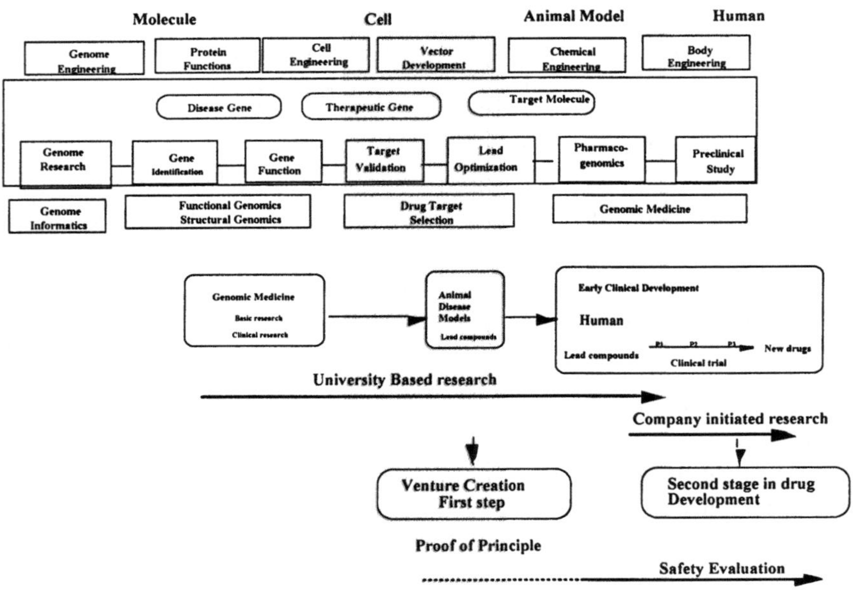

industry to the free zone or university to free zone. That is where the venture will flourish. This is a new type of academy-industry cooperation and the safety evaluation must be implemented into this. TLO (technology licensing organization) plus the supporting translational research center will bridge these two components.

Toxicogenomics in Genome Science World

These new areas in genome science world provide the basis for toxicogenomics. We can also create terminology such as toxico-Rnomis, toxico-Proteomics, etc. Pharmaco- and toxico-genomics can be considered in the context of the body's response to environmental stress. In view of extremely high cost necessary for drug development and evaluation of safety of new drugs, we have to develop intelligent approach enabling us to predict the behavior of genes, proteins, or drugs in the human body based on genomic information such as toxicogenomics in vitro at molecular, cellular, and body levels or toxicogenomics in silico collaborating with systems biology approach. Wealth of genome information made available for living organisms will open up the novel approach for "oriental medicine" such as Traditional Chinese Medicine (TCM:Figure 9).

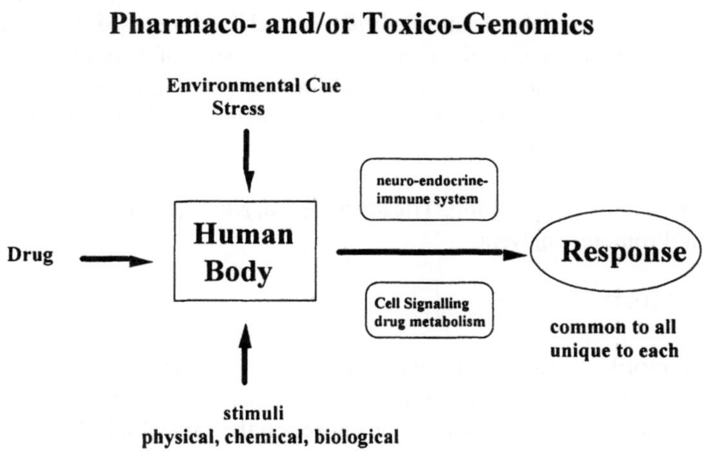

Fig. 9

Application of transcript profiling to mechanism-based toxicology

George Orphanides,* Jonathan G. Moggs, Tracy C. Murphy, John W. Edmunds, and William D. Pennie

Syngenta Central Toxicology Laboratory, Macclesfield, Cheshire, SK10 4TJ, UK
*Corresponding author. Email address: george.orphanides@syngenta.com

Summary. Technologies stemming from recent advances in the genomic sciences are set to revolutionise toxicology. Among these technologies is transcript profiling, which permits the simultaneous measurement of the expression levels of thousands of different genes. The application of transcript profiling to toxicology has been termed toxicogenomics and promises to accelerate identification of the molecular mechanisms through which xenobiotics induce adverse health effects. One of the main obstacles to the successful application of this technology is the lack of information on the relationship between the regulation of specific genes and the development of toxicity. In this article, we describe some of the approaches we are taking in our own laboratory to understand this relationship. Transgenic 'knock-out' mice and cell lines in which components of toxicant-signalling pathways have been modified or deleted are used to facilitate the identification of genes directly related to mechanisms of toxicity. In addition, analysis of gene regulatory regions can reveal the biological pathways that regulate transcriptional responses to toxicants, leading to a better understanding of molecular mechanisms of toxicity.

Key words. Transcript profiling, Transgenic mice, Modified cell lines, Gene regulation, Promoter analysis

Introduction

Transcript profiling allows the expression levels of thousands of genes to be measured simultaneously. The application of this technology to toxicology – toxicogenomics - is set to have a major impact on the toxicological sciences. This technique can be used in either "mechanistic" or "predictive" modes of analysis. In the mechanistic mode, transcript profiling is used to implicate specific genes or biological pathways in the mechanism of action of a toxicant. Such research leads are then subjected to further investigation using conventional techniques in order to clarify the roles, if any, they play in toxicity. Alternatively, this technology can be utilized in a predictive context, with the hope that biological responses induced by toxicants can be described based on comparison of global patterns of gene expression. In this way, the mode of action of a novel toxicant may be identified by comparing the expression pattern it elicits with established expression "fingerprints" of reference toxicants where mechanisms are understood. In this article, we discuss methods used in our own laboratory in the application of transcript profiling to mechanistic toxicology.

Application of transcript profiling to mechanistic toxicology: custom *ToxBlot* microarrays

The regulation of transcription is one of the principal mechanisms a cell uses to maintain homeostasis in a changing environment; therefore, most mechanisms of toxicity are accompanied by alterations in gene expression. Our laboratory has developed a series of microarrays that facilitate the transcriptional analysis of biological process relevant to mechanisms of toxicity (Pennie et al., 2001). Our *ToxBlot I* array is capable of measuring the expression of 600 human genes implicated in biological processes related to stress or toxic responses and provides a general overview of cellular functions. To facilitate more detailed analyses, we have developed *ToxBlot II*, which contains 12,500 elements and includes complete coverage of 199 functional categories of human genes, including all the major pathways represented on *ToxBlot I*. In addition, we have developed microarray platforms that permit the comprehensive analysis of gene expression events related to specific toxic endpoints, including endpoints that involve oxidative stress and perturbation of tyrosine catabolism.

How may a toxicant affect gene expression?

With the advent of transcript profiling, it is now possible to identify, rapidly and holistically, changes in the levels of gene transcripts associated with toxicant exposure. However, the increase in the rate at which gene expression data can be generated has not been accompanied by corresponding advances in our ability to interpret them as biologically meaningful information. In order to gain the full benefit of these technological advances, we must be able to translate gene expression data into phenotype (Smith, 2001). To do this we must first consider how a toxicant can affect gene expression. Figure 1 illustrates the flow of information from our genome - the DNA in our chromosomes - to an active protein. The genes in our genome are first transcribed into long molecules of RNA, which are then processed into messenger RNAs (mRNAs) before being translated into proteins. Following translation, many proteins then require further modification before they become active (e.g. phosphorylation, glycosylation or proteolytic cleavage). Normal cell function is achieved when this flow of information is unhindered. A toxicant can interfere directly with each of these steps to cause cellular dysfunction. This dysfunction may cause an adaptive response, as the cell attempts to compensate for the effects of the toxicant, and

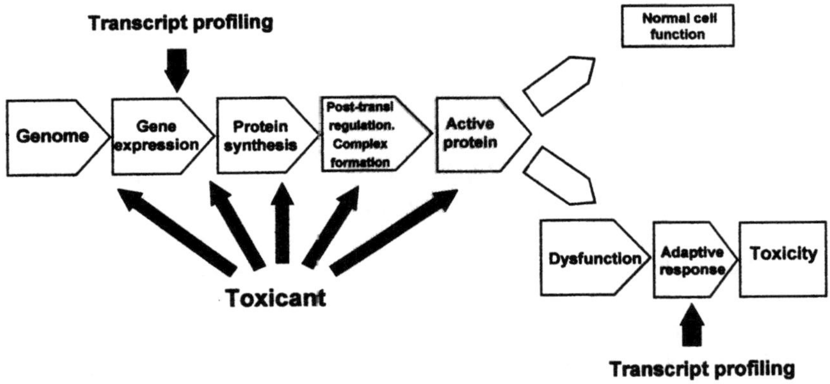

Fig. 1: How a toxicant affects gene expression. A toxicant can cause cellular dysfunction by interfering with the transfer of information from the genome to the production of an active protein. This dysfunction may then cause an adaptive response, as the cell attempts to compensate for the effects of the toxicant, and ultimately can result in toxicity. Transcript profiling measures changes in gene expression that occur as a direct consequence of toxicant exposure and changes associated with the adaptive response of the cell.

ultimately can result in toxicity to the organism. Transcript profiling measures changes in gene expression that occur as a direct consequence of toxicant exposure and changes associated with the adaptive response of the cell. These two types of toxicant-induced gene expression change are often referred to as "cause" and "effect" changes, respectively, and distinguishing these changes is a major challenge for toxicogenomics. Furthermore, most toxicants will induce the regulation of multiple genes, and only some of these genes will be related to the mechanism of toxicity. This can further complicate the interpretation of gene expression data.

Experimental design: the use of transgenic 'knock-out' mice and modified cell lines

Good experimental design can simplify the interpretation of transcript profiling data. This includes careful selection of the experimental system, compound dose and time of exposure, and the concurrent collection of conventional toxicology data (e.g. biochemical, clinical and histopathological data (Orphanides et al., 2002)). It is particularly important to acknowledge that changes in gene expression directly associated with the onset of toxicity will occur before a phenotypic change is seen. One approach that can facilitate the identification of gene expression changes directly linked to the molecular mechanism of a toxicant involves the use of animal models or cultured cell lines in which a pathway

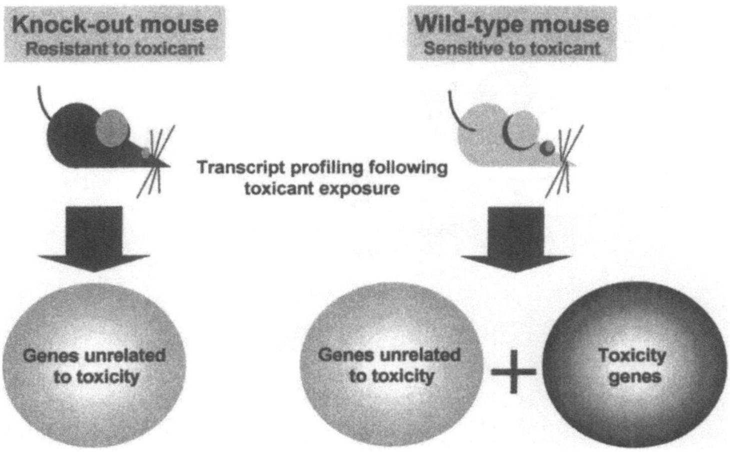

Fig. 2: The use of transgenic 'knock-out' mice and transcript profiling to identify genes associated with mechanisms of toxicity. Comparative transcript profiling of transgenic 'knock-out', mice resistant to the toxic effects of a xenobiotic, and wild type mice can be used to identify genes associated with compound toxicity. Alterations in gene expression seen in resistant knock-out mice dosed with a toxicant are unlikely to be linked to the adverse effects of the compound. Conversely, changes in gene expression that occur in sensitive wild-type animals, but not in resistant knock-out animals, are much more likely to be directly associated with the mechanism of toxicity.

relevant to the mode of action has been inactivated or modified. Transgenic 'knock-out' mice resistant to the toxic effects of a particular compound can be used to identify genes whose regulation is not directly related to the development of toxicity (Figure 2). For example, mice lacking active Peroxisome Proliferator Activated Receptor α are resistant to the adverse effects of peroxisome proliferators (PPs), a diverse group of non-genotoxic rodent hepatocarcinogens. (Roberts, 1998). Alterations in gene regulation seen in these knock-out mice exposed to toxicant are unlikely to be linked to the adverse effects of the compound. Any changes in gene expression that occur in a sensitive wild-type animal, but do not occur in a resistant knock-out animal, are more likely to be directly associated with the mechanism of toxicity. While, not all gene expression changes that match this description will be directly involved in the mode of action of a toxicant, this approach focuses attention on the most likely candidates. We are using this approach in our laboratory to identify genes involved the mechanism of toxicity of PPs.

A similar approach involves the use of cultured cell lines in which components of cellular pathways used by toxicants have been inactivated, activated or modified. We are using this approach to investigate the mechanism of action of endocrine disrupting chemicals (EDCs) that operate by activating the estrogen receptors (ERα and ERβ), nuclear receptors that function as ligand-dependent

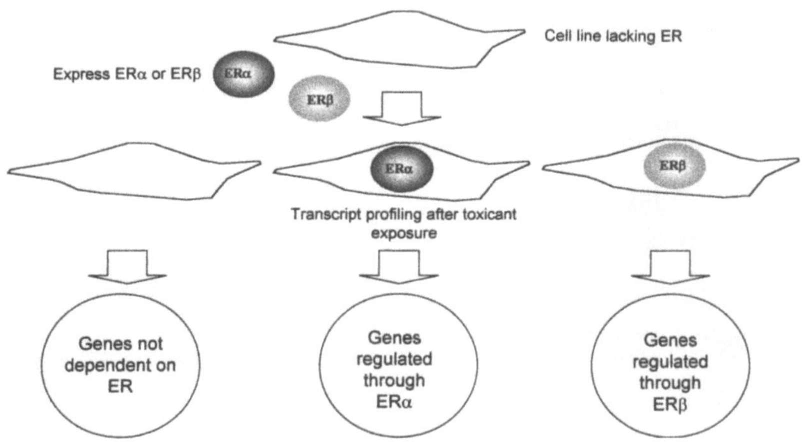

Fig. 3: Example of the use of cell lines modified to express components of toxicant signalling pathways to identify genes associated with toxicity. In this example, high-efficiency cell transfection of adenoviral expression constructs is used to express the two mammalian estrogen receptors (ERα and ERβ) in a human breast cell line that lacks functional ERs. Cells expressing ERs, and control cells lacking receptor, are then treated with endocrine disrupting compounds that target the ERs and their transcript profiles are measured. This approach facilitates the identification of genes regulated through ERα and ERβ, and genes regulated via mechanisms that are independent of these receptors.

transcription factors (Moggs and Orphanides, 2001). The two receptors are differentially targeted by EDCs, with the phytoestrogens genistein and coumestrol having a much higher affinity for ERβ (Kuiper, 1998). We aim to identify genes regulated by different EDCs through the two ERs. The two receptors appear to have overlapping, but distinct, gene targets, suggesting that the regulation of ERα and ERβ by EDCs will result in different biological responses (Pennie at al., 1998; Routledge et al., 2000). To identify the genes induced by different EDCs through each receptor, we have expressed either ERα or ERβ in a human breast cell line that lacks functional ERs. Expression of the receptors is achieved by high-efficiency cell transfection of adenoviral expression constructs (He et al., 1998). Cells expressing ERs, and control cells lacking receptor, are treated with compounds and their transcript profiles are measured. This approach facilitates the identification of genes regulated through ERα and ERβ and genes regulated via mechanisms that are independent of the receptors (Figure 3). The general principle of this cell culture-based approach will have application in many areas of mechanistic toxicology.

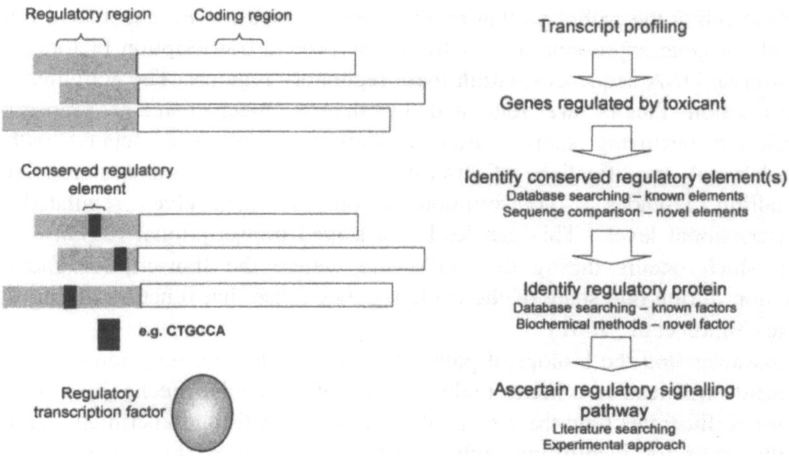

Fig. 4: Sequence analysis of gene regulatory regions as the basis for characterising the biological pathways that regulate transcriptional responses to toxicants. The regulatory regions of genes induced or repressed by toxicants are first identified with the aid of promoter databases or genome sequence browsers. Genes co-ordinately regulated through the same pathway, and therefore by the same transcription factor(s), are predicted to contain the same DNA sequence motifs in their regulatory regions. These conserved motifs can be identified by comparing the regulatory sequences of co-ordinately regulated genes. The transcription factor(s) that bind to these conserved DNA motifs to regulate gene expression can then be identified. The identity of the controlling transcription factor(s) may be sufficient to reveal the pathway of gene regulation. Where this is not known, or where the transcription factor has not been characterised previously, this may be determined by further experimentation.

Understanding why genes are regulated upon toxicant exposure through the analysis of regulatory regions

The identification of genes regulated by exposure to toxicants is now relatively easy. However, understanding the cellular pathways that control the expression of these genes is much more challenging. For example, it has long been known that treatment of mice with the toxicant pregnenolone-16α-carbonitrile (PCN) leads to hepatic induction of the genes encoding cytochrome P450 3A enzymes (CYP3As). However, it was discovered only recently that this is due to the activation by PCN of the Steroid and Xenobiotic Receptor (SXR), which binds to the regulatory regions of the CYP3A genes (Blumberg et al., 1998; Kliewer et al., 1998). The availability of genomic sequences and web-based genome sequence browsers (for a review of genome sequence browsers, see http://www.sciencemag.org/feature/plus/sfg/resources/res_maps.shtml) has simplified the identification of putative gene regulatory regions. The sequences of these regulatory regions are rich in information, and hold the key to understanding the pathways that regulate gene expression in response to toxicants. Levels of gene expression are controlled by protein transcription factors that bind to discrete DNA sequences within these regulatory regions. The activities of these transcription factors are regulated by diverse mechanisms. Transcriptional regulation occurring shortly after a stimulus is often associated with post-translational modification of transcription factors by kinases regulated by signalling cascades. Transcription factors are themselves regulated at the transcriptional level. This can lead to a staged transcriptional response, such as that which occurs during the cell cycle, where the transcription factors that function during one stage of the cycle regulate those that function during the next stage (Simon et al., 2001).

Characterising the biological pathways that regulate transcriptional responses to toxicants will lead to a better understanding of molecular mechanisms of toxicity. Figure 4 illustrates how the results of a transcript profiling experiment can be used as the basis for identifying cellular pathways perturbed by a xenobiotic. The regulatory regions of genes induced or repressed by a toxicant are first identified with the aid of promoter databases (e.g. http://www.epd.isb-sib.ch/) or genome sequence browsers (e.g. http://www.ncbi.nlm.nih.gov/genome/guide/human/). Genes co-ordinately regulated through the same pathway, and therefore by the same transcription factor(s), are predicted to contain the same DNA sequence motifs in their regulatory regions. These conserved motifs can be identified by comparing the regulatory sequences of co-ordinately regulated genes. The transcription factor(s) that bind to these conserved DNA motifs to regulate gene expression can then be identified. This is accomplished by searching transcription factor databases for proteins that bind to the sequence identified (e.g. http://transfac.gbf.de/TRANSFAC/), or experimentally, using biochemical methods (e.g. Kadonaga and Tjian, 1986). The identity of the controlling transcription factor may be sufficient to reveal the pathway of gene regulation. Where this is not known, or where the transcription factor has not been characterised previously, this must be determined by further experimentation. In order to illustrate the principle of this method, we have described a simplified scenario in which a single transcription factor controls the expression of a set of co-ordinately regulated genes. However, it is important to consider that many genes are regulated by the concerted action of multiple transcription factors. This approach has been used to identify a novel regulatory element required for circadian control of gene expression in *Arabidopsis thaliana* (Harmer et al., 2001).

Summary

Successful application of transcript profiling to mechanistic toxicology requires an understanding of the link between gene expression changes and phenotype. In this article, we have described how the use of transgenic mice and modified cell lines can help us to distinguish genes directly associated with mechanisms of toxicity from genes that are regulated inconsequentially, or through mechanisms unrelated to toxicity. We must also make full use of the information presented to us by genome sequencing programmes; not only the identity and sequences of genes, but also the sequences of gene regulatory regions. As we have described, these regulatory regions define why particular groups of genes are induced or repressed by a xenobiotic, and their analysis will provide insight into the molecular alterations that occur upon toxicant exposure. The development of methods that define the link between genes and phenotype must be given equal priority to the development of techniques that measure gene expression levels. Only then will the rate at which we can interpret transcript profiling data approach the rate at which we can generate them. This will be crucial for the successful application of toxicogenomics to mechanistic toxicology and risk assessment.

Acknowledgements

We thank Dr. Ian Kimber for comments on the manuscript. Part of the work described in this article is funded by the Food Standards Agency.

References

Blumberg B, Sabbagh W Jr, Juguilon H, Bolado J Jr, van Meter CM, Ong ES, Evans RM (1998) SXR, a novel steroid and xenobiotic-sensing nuclear receptor. Genes Dev 12: 3195-3205

Harmer SL, Hogenesch JB, Straume M, Chang HS, Han B, Zhu T, Wang X, Kreps JA, Kay SA (2000) Orchestrated transcription of key pathways in *Arabidopsis* by the circadian clock. Science 290:2110-2113.

He TC, Zhou S, da Costa LT, Yu J, Kinzler KW, Vogelstein B (1998) A simplified system for generating recombinant adenovirus. Proc Natl Acad Sci 95:2509-2514

Kadonaga JT, Tjian R (1986) Affinity purification of sequence-specific DNA binding proteins. Proc Natl Acad Sci 83:5889-5893

Kliewer SA, Moore JT, Wade L, Staudinger JL, Watson MA, Jones SA, McKee DD, Oliver BB, Willson TM, Zetterstrom RH, Perlmann T, Lehmann JM (1998) An orphan nuclear receptor activated by pregnanes defines a novel steroid signaling pathway. Cell 92: 73-82

Kuiper GG, Lemmen JG, Carlsson B, Corton JC, Safe SH, van der Saag PT, van der Burg B, Gustafsson JA (1998) Interaction of estrogenic chemicals and phytoestrogens with estrogen receptor beta. Endocrinol 139:4252-4263

Moggs JG, Orphanides G (2001) Estrogen receptors: orchestrators of pleiotropic cellular responses. EMBO Rep 2:775-781

Orphanides G, Pennie WD, Moffat GJ Kimber I (2002) Toxicogenomics: theoretical and practical considerations. Comments on Tox, in press

Pennie WD, Aldridge TC, Brooks AN (1998) Differential activation by xenoestrogens of ER alpha and ER beta when linked to different response elements. J Endocrinol 158:R11-14

Pennie WD, Woodyatt NJ, Aldridge TC, Orphanides G (2001) Application of genomics to the definition of the molecular basis for toxicity. Toxicol Lett 120:353-358

Roberts R (1999) Peroxisome proliferators: mechanisms of adverse rodent effects and molecular basis for species differences. Arch Toxicol 73:413-418

Routledge EJ, White R, Parker MG, Sumpter JP (2000) Differential effects of xenoestrogens on coactivator recruitment by estrogen receptor (ER) α and β. J Biol Chem 275:35986-35993

Simon I, Barnett J, Hannett N, Harbison CT, Rinaldi NJ, Volkert TL, Wyrick JJ, Zeitlinger J, Gifford DK, Jaakkola TS, Young RA (2001) Serial regulation of transcriptional regulators in the yeast cell cycle. Cell 106:697-708

Smith LL (2001) Key challenges for toxicologists in the 21st century. Trends Pharmacol Sci 22:281-285

Strategy for Toxicogenomics

Application of DNA microarrays for predicting toxicity and evaluating cross-species extrapolation

Russell S. Thomas,[1,2] Kevin R. Hayes,[1] Gina M. Zastrow,[1] Karen Tran,[2] Sharron G. Penn,[2] David R. Rank,[2] and Christopher A. Bradfield[1]

[1]McArdle Laboratory for Cancer Research, University of Wisconsin Medical School, Madison, Wisconsin 53706
[2]Aeomica, Sunnyvale, California 94085, USA

Summary. The application of DNA microarray technology to the field toxicology has increased significantly. In most cases, the research has monitored global changes in gene expression in order to provide insight into the cellular mechanisms of toxicity. Although assessing global gene expression changes may prove to be important when characterizing the action of a particular chemical, it is not necessarily predictive of the toxicological behavior within an organism or across species. As a first step for developing predictive toxicological models within a species, we developed a statistical model to classify 24 model treatments that fall into five well-studied toxicological categories based on gene expression. Using all the gene expression measurements resulted in relatively poor predictive accuracy. However, focusing on a diagnostic subset of genes greatly increased the predictive accuracy. For evaluating the toxicological behavior across species, a set of orthologous microarrays were developed that allowed a direct comparison of gene expression changes in both organisms. To construct these microarrays, a genome wide comparison of available human and mouse sequence was performed to identify putative orthologous genes. A subset of the orthologous genes were spotted on tandem microarrays (one human and one mouse) and used to evaluate conservation of expression patterns between organisms. The use of predictive statistical models and cross-genome comparisons in chemically induced gene expression are the next logical advancements in the field of toxicogenomics and their application has the potential to be extremely valuable in regulatory decisions and the risk assessment process.

Keywords. Genomics, Toxicology, Toxicogenomics, Microarrays, Species Extrapolation

---

Correspondence should be addressed to: Russell S. Thomas, Aeomica, 928 E. Arques Ave., Mail Stop #9, Sunnyvale, CA 94085
Email: Rusty.Thomas@am.amershambiosciences.com,
Voice: (408) 737-4208, Fax: (408) 737-4891

Introduction

Due to the recent advances in genomic technology, measuring gene expression changes is quickly becoming an integral part of characterizing the effects of chemical exposure on a particular organism. The underlying belief is that identifying gene expression changes will pinpoint the disrupted cellular and molecular pathways that lead to toxicity. However, just identifying the disrupted pathways and associated gene expression changes do not necessarily provide a method to predict similar toxic responses with other chemicals or across species. A big challenge for the emerging field of toxicogenomics will be to develop models and tools that use gene expression measurements to ultimately predict toxicity in untested chemicals and also determine whether a similar toxic response will occur in humans.

Application of genomics to toxicity testing

Under current regulatory polices, chemicals that are believed to comprise both a significant exposure and heath risk in the human population are selected to undergo testing for toxicity and carcinogenic potency. The traditional method of evaluating carcinogenic activity and chronic toxicity of a specific chemical has been the animal bioassay. The typical experimental design for these studies is two years in duration and involves animals of both sexes and usually from one strain of rat and/or mice. Several dosage levels are chosen and detailed records on animal health, vital statistics, and pathology are kept throughout the study. Due to the amount of resources involved, each bioassay costs between 2 and 4 million dollars and takes several years to complete (NTP 1996). According to the National Toxicology Program (NTP), the number of chemicals currently tested by the NTP stands at 505 in long-term studies, 66 in short-term tests,. and only a single sub-chronic study (http://ntp-server.niehs.nih.gov/). By comparing the number of completed bioassays with the 70,000 to 85,000 chemicals in commerce today (Ziegler 1993; Lucier and Schecter 1998), it is clearly impossible to apply the current testing system to all chemicals of concern (Fig. 1). Alternative approaches must be developed in the interest of protecting human health and saving economic resources.

The development of an efficient test that would allow prioritization of untested chemicals based upon their toxic potential would have a significant impact on how efficiently we evaluate chemicals of concern. One approach for predicting toxic potential is to classify chemicals based upon their capacity to alter gene expression in a manner that is similar to known toxicants or chemicals previously tested in the two-

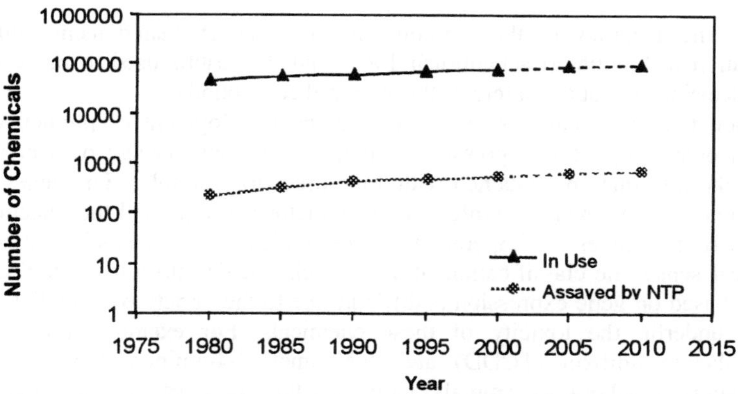

Fig 1. A comparison between the estimated number of chemicals in use and the number of chemicals tested in chronic animal bioassays by the National Toxicology Program (NTP). Estimates of the number of chemicals in use were derived from previously published reports (Ziegler 1993) and the number of chemicals tested was obtained from the NTP web site (http://ntp-server.niehs.nih.gov/). Based on the current rate of testing, it is apparent that the number of chemicals tested will never approach the number of chemicals being used.

year bioassay. Chemicals that produce gene expression patterns similar to those of established toxicant could be classified and regulated as potentially toxic pending a more thorough examination. By taking this approach, one would have to make the assumption that all toxic chemicals will alter gene expression at some level. Given that toxicity results from some form of cellular dysfunction or death, gene expression changes are likely to occur at some level. Regardless of whether the expression changes are associated with the root cause of the toxicity or downstream of the initial event, the resulting overall pattern of gene expression changes can act as a diagnostic pattern for that chemical to compare with other known toxicants.

The widespread application of microarray technology to toxicity testing would require the integration of sophisticated statistical tools into the analysis. Specifically, the use of statistical classification techniques to build predictive models for the subsequent classification of unknown/untested chemicals. These statistical techniques range from linear and non-linear discriminant analysis (Huberty 1994) to nearest-neighbor approaches (Dasarathy 1991) and neural networks (Fausett 1994). With any of these techniques, a training dataset is needed where the model is trained on data where the answer is already known. For example, a potentially beneficial dataset would be to use a subset of the compounds already tested by the NTP in their bioassay. These compounds could be fed to small numbers of rodents in short-term studies and the changes in gene expression evaluated using microarray technology. Provided that the subset of chemicals contains chemicals both positive and negative in the bioassay, a classification model could be built to predict the response in subsequent untested chemicals. The accuracy of these predictions could be evaluated using additional cross-validation techniques (Lachenbruch 1967) and the results used to make critical regulatory decisions about the safety of the analyzed compounds.

There are two important points concerning the development of predictive toxicological models using gene expression. First, the ultimate success of a model dependent on the information contained within the predictor variables (i.e., genes on the array). Second, the removal of uninformative predictor variables and the selection of a diagnostic subset of genes are imperative for both biological and statistical reasons. In the biological sense, the classification of a set of chemicals into a toxicological class or endpoint based on gene expression is difficult due to the variety of potential mechanisms that underlie the toxicity of these chemicals. For example, both 2,3,7,8-tetrachlorodibenzo-*p*-dioxin (TCDD) and *N,N*-dimethyl-4-aminoazobenzene (DAB) are carcinogenic in laboratory animals. However, they produce the same toxic endpoint, cancer, through different mechanisms. Dioxin is a nongenotoxic carcinogen that acts through a ligand-activated transcription factor known as the Ah receptor while DAB is metabolized to an active form that is mutagenic. Even though the toxicological endpoint may be the same, these two chemicals will undoubtedly activate and repress different biochemical and molecular pathways leading to different global patterns in gene expression. Therefore, in order to construct a predictive model for chemically induced cancer, the global gene expression dataset must be reduced to a subset of genes that are diagnostic across all chemicals.

From a statistical perspective, selection of a subset of diagnostic genes is also important. Statistical models with few predictors relative to the overall sample size typically yield more accurate (i.e., less biased) and more precise estimators (Hora and Wilcox 1982; Huberty 1994). For Bayesian models, underlying assumptions are made about the dependency of the variables. If the assumptions are violated in the full data set, better classification results can be obtained by removing uninformative predictor variables making the dependency assumptions less severe.

The widespread application of microarray analysis and statistical classification tools to chronic toxicity studies has the potential to be extremely useful from both a scientific and economic perspective. In general, the application would allow the construction and validation of a statistical models based on gene expression patterns that would predict the long-term toxicity of an unknown chemical based on short-term rodent studies using only a small number of animals. The potential cost and timesavings could be immense and the current gap between the number of tested and untested chemicals could be reduced in the interest of public health. However, there are significant hurdles to overcome before this type of analysis will be useful on a large scale. First, multiple factors converge to ultimately influence the manifestation of toxicity and the associated gene expression patterns. Among these factors are time, dose, route of administration, age of the animal and sex. Fully characterizing the influence of all of these variables on transcript profiles with even a small number of treatments would require considerable resources (e.g., 200 treatments * 3 time points per treatment * 3 doses per time point * 3 routes per dose * 3 ages per treatment * 2 sexes = 32,400 microarray studies). Conceptually, some of these variables could be standardized in the animal studies and, in principal, the classification models incorporate a number of these variables since it statistically relates gene expression measurements performed in a short term study to the results from the chronic bioassay. Whether or not, the classification models will be robust enough to allow accurate predictions despite these factors remains to be seen. Second, the training set must be carefully chosen to incorporate a broad spectrum of toxicants that act via different mechanisms. The ability of the statistical model to predict long term toxicity of a chemical that differs significantly from the training set will be limited. Therefore, the larger the public database of chemically induced gene expression and the more toxicological categories added to the model, the more predictive the models will become. In the future, many of these problems will be overcome and the era of the resource-intensive animal bioassay will end.

Application of genomics to species extrapolation

It is well established that the majority of known human chemical carcinogens are also carcinogenic in at least one species of laboratory animal. Whether the reverse is true is difficult to establish with any degree of confidence. But, the assumption that it is true provides part of the underlying foundation for toxicological testing in animal systems and is the basis for one of the underlying tenets in toxicology- that experimental results in animal models, given certain restrictions, are generally applicable to humans. Regulatory agencies, academic researchers, pharmaceutical companies, and chemical companies are all heavily reliant on results from animal studies for establishing health and safety guidelines, assessing toxicity, and evaluating the potential efficacy of new drugs. For example, animal studies are commonly used at the Environmental Protection Agency (EPA) in both the hazard identification and response assessment phases of the risk assessment process. During hazard evaluation, the EPA evaluates a chemical's toxicological profile using laboratory animal models (i.e., the type and extent of harmful effects) while the response assessment includes additional dose response and chronic exposures in order to find the "No-Observed-Effect-Level" (NOEL) and calculate a "Reference Dose" (RfD). Although additional safety factors are built into calculating the reference dose, the basis for its calculation relies on results from the animal studies.

In spite of the heavy reliance on animal models in toxicological research, most toxicologists recognize that significant quantitative and qualitative differences exist between humans and the animal models. For example, there are significant differences in the responses of rodents and humans to peroxisome proliferators (Vanden Heuvel 1999) and the LD_{50} for TCDD varies by more than 1000-fold between species (Poland and Knutson 1982). With these examples and many others, it is evident that

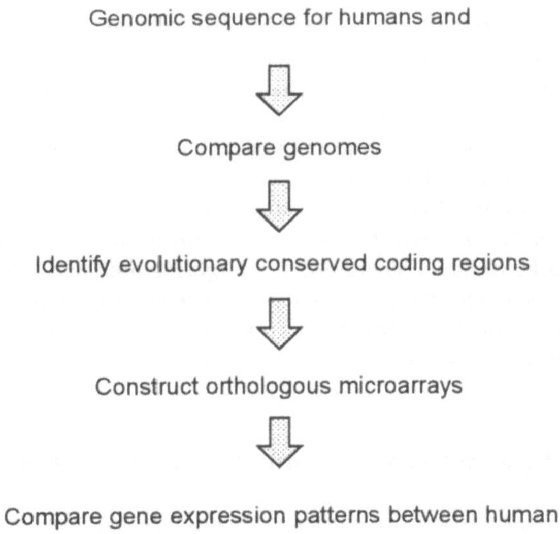

Fig 2. A flow chart describing the construction of orthologous microarrays to address species extrapolation issues in toxicology. From the top of the flow chart, the available genome sequences of both humans and the important rodent models are used to identify evolutionary conserved regions. These regions are screened for coding potential and arrayed on glass slides. Measurements of gene expression in both organisms following chemical insult allow the assessment of conserved toxicological endpoints at the molecular level.

the identification of toxicological species differences and their implementation in the regulatory arena are essential.

The development of a system that would allow the assessment of all molecular differences between species following chemical exposure would have a significant impact on how we evaluate the pharmacodynamic aspects of cross-species extrapolations. Although assessing all molecular differences is nearly impossible, one potential approach would be to measure expression changes in orthologous sets of genes (Fig. 2). Orthologs are genes in different species that evolved from a common ancestral gene by speciation. Test chemicals that induce different transcriptional responses between species could be classified as harboring potential pharmacodynamic differences and put through to additional mechanistic studies and regulatory review. By taking this approach, one would have to make two assumptions. First, for many chemicals and cell types, in vitro results from human tissues would have to be similar to those obtained in vivo due to obvious limitations on human studies. Second, conserved changes in gene expression equate to conserved pharmacodynamic endpoints.

For the first point, arguments for and against extrapolating from in vitro results to in vivo are important but will not be addressed here. In support of the second assumption, evolution and selection has maintained the structure and function of many biochemical pathways over time resulting in the conservation of many important processes (Schmidt-Nielsen 1975). Although the function of some genes may have changed over time, similar changes in orthologous sets of genes are likely to produce the same physiological and toxicological endpoint whether it is inflammation, proliferation, apoptosis, necrosis, or cellular differentiation.

In order to develop a system that can exhaustively measure alterations in orthologous genes, the majority of the genomic sequence of both species must be known and the sequences compared to identify evolutionarily conserved coding regions. For toxicology, this would mean the completion of the mouse (currently at 15% in both draft and finished BAC sequence plus approximately 3X coverage in whole genome shotgun sequence), rat (currently at 7% in both draft and finished BAC sequence plus approximately 1.5X coverage in whole genome shotgun sequence), and human genomes (currently at 98.5% in both draft and finished sequence) (http://www.ncbi.nlm.nih.gov/Genomes/index.html). Already, previous studies have compared large segments (> 30 kb) of human and rodent sequences demonstrating that coding domains are generally well conserved while non-coding regions exhibit variable levels of conservation (Koop 1995). After completion of the genomes, the sequences could be compared using various sequence alignment algorithms and the results screened for coding potential (i.e., containing an open reading frame over the length of the conserved sequence). The resulting list would contain orthologous sets of putative exons that could be used to create parallel rodent and human microarrays.

Fundamentally, the issue of cross-species extrapolation is at the root of both biological and toxicological research and the subsequent regulatory decisions. The application of genomics to species extrapolation issues in the form of cross-species sequence comparisons and orthologous microarray analysis has the potential to directly address this issue and provide critical molecular data on the way different species react to toxicants. This information would allow the integration of both pharmacokinetic and pharmacodynamic data at a basic level and significantly improve the uncertainty associated with conventional risk assessments.

Conclusions

Although the study of genes relative to chemical toxicity has been around for a number of years, the recent progress in genome sequencing and the ability to simultaneously monitor the expression of thousands of genes have moved genomics to the forefront of toxicology and spurred the coining of the term toxicogenomics. The near completion of the human genome and the progress on the rodent sequencing marks a new era, not just in biology, but in toxicology as well. But, the availability of the sequence information and the development of the new genomic tools are only part of the effort. The integration of this information and application to problems such as toxicity testing and cross-species extrapolation are the goals of the future and where some of the true benefits of these advancements lie.

References

Dasarathy BV (1991) Nearest Neighbor (NN) Norms: NN Pattern Classification Techniques. IEEE Computer Society Press, Los Alamitos, CA

Fausett L (1994) Fundamentals of Neural Networks: Architectures, Algorithms, and Applications. Prentice Hall, Englewood Cliffs, NJ

Hora SC, Wilcox JB (1982) Estimation of error rates in several population discriminant analysis. J Marketing Res 19:57-61.

Huberty CJ (1994) Applied Discriminant Analysis. John Wiley, New York.

Koop BF (1995) Human and rodent DNA sequence comparisons: A mosaic model of genomic evolution. Trends Genet 11:367-371

Lachenbruch PA (1967) An almost unbiased method of obtaining confidence intervals for the probability of misclassification in discriminant analysis. Biometrics 23:639-645

Lucier GW, Schecter A (1998) Human exposure assessment and the National Toxicology Program. Environ Hlth Perspect 106:623-627

NTP (1996) National Toxicology Program: Annual Plan for Fiscal Year 1996. National Institutes of Health, Department of Health and Human Services, Washington, DC.

Poland A, Knutson JC (1982) 2,3,7,8-Tetrachlorodibenzo-p-dioxin and related halogenated aromatic hydrocarbons: examination of the mechanism of toxicity. Ann Rev Pharmacol Toxicol 22:517-554

Schmidt-Nielsen K (1975) Scaling in biology: the consequences of size. J Exp Zoology 194:287-307

Vanden Heuvel JP (1999) Peroxisome proliferator-activated receptors (PPARs) and carcinogenesis. Toxicol Sci 47:1-8

Ziegler J. (1993) Health risk assessment research: The OTA report. Environ Hlth Perspect

Use of reverse genetics and cDNA arrays to understand 'dioxin' toxicity

Andrew G. Smith, Susan Robinson, Bruce Clothier, Reginald Davies, David J. Judah, Joan Riley, and Timothy W. Gant

MRC Toxicology Unit, Hodgkin Building, Leicester University, Leicester LE1 9HN, UK

Summary. The molecular mechanisms of toxic agents are rarely likely to be the result of change in the expression of a single gene. Even apparently simple actions of chemicals are probably the consequence of genetically variable multigene expression. One strategy for attacking this problem is to take advantage of the genetic variation of response in mice to search for susceptibility genes using genetic linkage analysis in combination with comparisons of gene expression in the parent strains by cDNA microarray technology. Genetic variation in mice and toxicogenomics were used to explore mechanisms of gene interaction leading to cell malfunction and injury in the liver caused by dioxin. This demonstrated susceptibility loci, other than the *Ahr* gene, pertinent to the development of porphyria (a disruption of heme synthesis) and liver injury. cDNA arrays of 4000 IMAGE clones pertinent to toxicology were used to compare candidate multiple gene expression in strains relative to their initial hepatic response e.g. induction of drug metabolism enzymes, and to their subsequent development of porphyria and liver injury. Phenotypic response was compared with gene expression by metabolic system including groups of genes for heme and iron metabolism, the AH battery and oxidative stress.

Key words. Porphyria, Liver injury, Dioxin, Susceptibility loci, Toxicogenomics

Genetic variation and toxicogenomics

Background

Pathological change, from whatever instigation, is always the end point following complex series of molecular events. Indeed the complexity involved is only now being fully realized with the introduction of new techniques for studying multigene expression and multiple protein profiles. All of these events have the potential for modulation by intrinsic genetic factors such as variants of oncogenes or promoter sequences of genes concerned with drug metabolism. The rapid progress in the development of genomics and proteomics has the possibility of detecting much greater molecular subtleties than was previously possible. The variable response to therapy of such diseases such as non-Hodgkin's lymphoma has proven to be due to molecular heterogeneity in the tumors not recognized until the advent of gene expression profiling (Alizadeh et al. 2000). Similarly, the response of individuals to toxic chemicals is likely to be dependent on greater molecular variation than we have previously anticipated. Although polymorphisms in some drug metabolism enzymes are well established (Meyer and Zanger 1997; Oscarson et al. 2001) there are likely to be many genetically variable factors, in say cell signalling pathways or oxidative stress responses, that we have not yet identified but are important in the mechanisms of action of chemicals.

Use of genetic variation to study mechanisms

The genetic variation between individuals that influences their responses to drugs or toxic chemicals can be used for investigative purposes as a tool to understand molecular mechanisms. One of the classic examples of this is the important role of inbred mouse strains to elucidate the mechanism of action of ligands for the aryl hydrocarbon (AH) receptor (Nebert et al. 1993). Not only did this lead to the demonstration of the control of expression of CYP1A isoforms and illustrate an important receptor polymorphism potentially present in humans, but also opened up avenues of interactions with other transcription factors such as ARNT and HIF1α that had not previously been expected from classical studies of toxicology (Gu et al. 2000). More recently, the development of 'knockout' mice has led to rapid advances in our standing of the importance of particular genes in normal and pathological gene function. However, it is important to remember that the use of this technology will usually tell us about the essentiality of a particular gene product but not about the profound influence of variability in the expression of other genes. For instance, although the 'knockout' of the mouse equivalent of the human hemochomatosis gene confirmed the crucial role in controlling iron absorption, it has soon become clear from inbred mouse studies that other unidentified genes are important for determining penetrance (Flemming et al. 2001).

Genetic mapping of offspring from crosses of significantly differently responding parent lines can be used to identify these other susceptibility loci, so-called reverse genetics. In toxicology it should also be possible to integrate these types of studies with the advances in toxicogenomics to gain insights into the influence of susceptibility on multiple gene expression and thus on fundamental molecular mechanisms (Fig. 1).

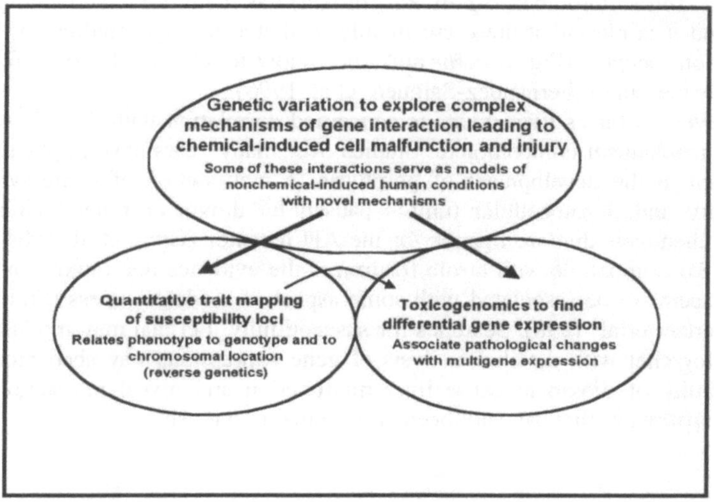

Fig. 1. Strategy to combine susceptibility studies with toxicogenomics to elucidate mechanisms of toxicity.

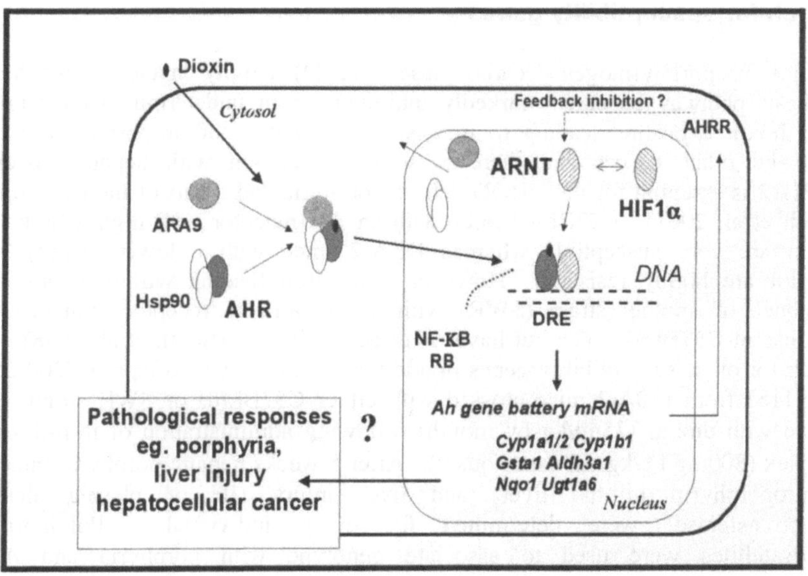

Fig. 2. Scheme of intracellular action of dioxin via the AH receptor for hepatic response. The link between gene expression and pathological change is still unclear.

2,3,7,8-Tetrachlorodibenzo-*p*-dioxin (dioxin) has many toxic and carcinogenic actions and it is clear that these are mainly mediated through binding to the aryl hydrocarbon receptor (Fig 2). *Ahr* null mice show no effects of dioxin for those parameters measured (Fernandez-Salguero et al. 1996).

However, as far as liver injury is concerned correlation with the AH receptor mediated mechanism is incomplete. Studies over many years have implicated iron metabolism in the development of porphyria (a malfunction of heme synthesis) liver injury and hepatocellular tumors caused by dioxin or related chlorinated aromatic chemicals that are ligands for the AH receptor (Greig et al. 1984; Smith et al. 1998). Compatible with a role for iron is the evidence that dioxin toxicity in the liver seems to be associated with some aspect of oxidative stress (Smith et al. 1998; Shertzer et al. 1998). Searches for susceptibility loci that may modulate *Ahr* function together with detailed analysis of gene expression, may shed more light on the ability of dioxin to cause liver malfunction and reveal interactions with metabolic pathways that have not been appreciated previously.

Multigene interaction in dioxin toxicity

Search for susceptibility genes

In mice, uroporphyrinogen decarboxylase (UROD) activity in the hepatic heme synthesis pathway becomes markedly inhibited by an undetermined mechanism after dioxin exposure, leading to massive accumulation of uroporphyrin in the liver and other tissues and there is some association with hepatic toxicity. CYP1A2 is essential for the UROD defect, porphyria and a part of the liver injury (Smith et al. 2001). C57BL/6J mice with an AH receptor with high affinity for dioxin are very susceptible whereas DBA/2 mice with a lower affinity AH receptor are highly resistant. However, prior iron loading will overcome the resistance of another strain (SWR) with a low affinity receptor, enhance the response in C57BL/6J mice but have no effect in DBA/2 (Smith et al. 1998). To search for other susceptibility genes in addition to the *Ahr*, F_2 offspring (200-250) were bred from DBA/2 mice crossed with either C57BL/6J or SWR strains and treated with dioxin (75μg/kg by mouth) following administration of iron-dextran complex (800mg Fe/kg subcutaneously). After 5 weeks hepatic porphyria (nmol/g of uroporphyrin in the liver) and liver injury (IU of plasma alanine aminotransferase) were determined for every individual. Polymorphic microsatellites were used to associate genotype with porphyria and ALT phenotypes. High and low responders in both crosses were genotyped for two microsatellites per chromosome. Where the preliminary findings suggested the presence of a susceptibility locus all mice were genotyped with up to 12 markers per target chromosome. Besides the polymorphism at the *Ahr* gene in the C57xDBA/2 cross, susceptibility loci (quantitative trait loci) were found on chromosomes 1,9, 11 and 14 depending on the phenotype and cross (Fig.3). There are candidates for some of these loci but attention is being paid particularly to comparison with gene expression differences in the parent strains that might reflect the susceptibility loci and the underlying mechanisms.

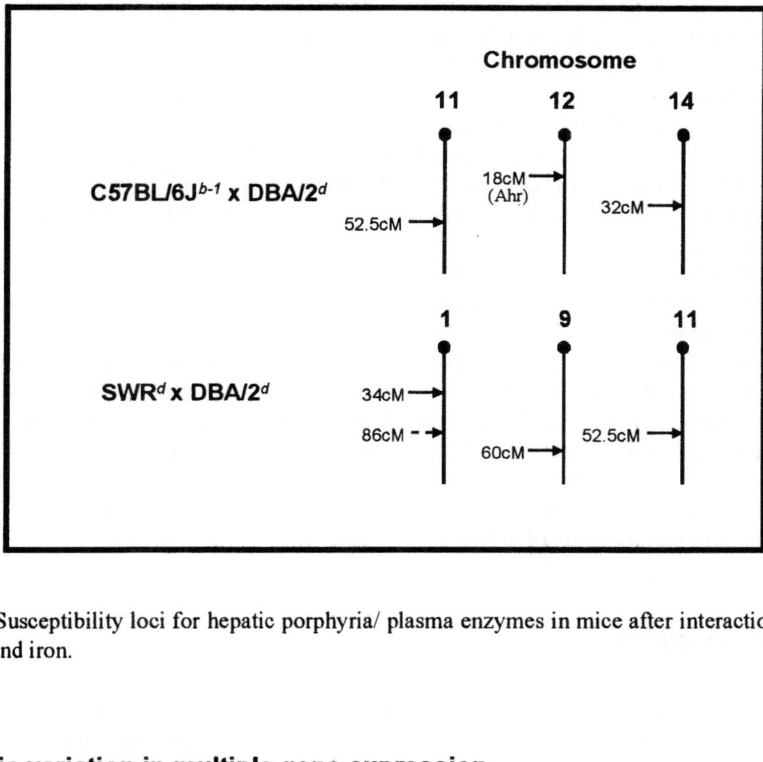

Fig. 3. Susceptibility loci for hepatic porphyria/ plasma enzymes in mice after interaction of dioxin and iron.

Genetic variation in multiple gene expression

As a complementary approach to the susceptibility loci searches, we investigated the expression of many genes that might be anticipated to be associated with porphyria and liver damage. The porphyric response of mice following a single dioxin dose can take up to a 5 weeks to develop (Smith et al. 1998). Clearly the initial abilities of dioxin to induce gene expression within a few days may be different to that seen as a consequence of mechanisms producing toxicity. Thus in studies of this sort it is important to follow gene expression over a time period compared with the phenotype. cDNA microarrays on glass were constructed from approximately 4000 mouse IMAGE clones for candidate genes associated with drug metabolism, oxidative stress, heme synthesis, iron metabolism, apoptosis and general liver injury as described for a human collection (Turton et al. 2001). It was clear that after 2 or 5 weeks following exposure many genes were up or down regulated by dioxin in the liver of C57BL/6J mice but little at all in the DBA/2 strain. These included genes associated with oxidative stress related processes, for instance a significant (5-10 fold) down regulation of *Car3* as observed in rats for 3,3',4,4',5-pentachlorobiphenyl (Ikeda et al. 2000), and an up regulation of the antioxidant transcription factor gene *Nfe2l2* and collagen synthesis genes. These changes were not observed 3 days following dioxin and were obviously occurring in response to a genetically variable fundamental toxic process perhaps of an oxidative nature. Interestingly, analysis of those genes classically considered to be controlled in the AH battery (Nebert et al. 1993) showed differential expression between the strains only for some and not others. *Cyp1a1* and *Cyp1a2* were massively up regulated (>30 fold) at 3 days in both high and low affinity strains for the AH receptor and this still persisted at 5 weeks (>12 fold in both C57BL/6J and DBA/2 mice). In contrast, *Dia4* (diaphorase) was only induced in the susceptible C57BL/6J strain. Porphyria is due to a malfunction of heme synthesis. Expression of the rate controlling enzyme *Alas1* was only slightly induced after 3 days despite the large initial induction of CYP isoforms in both strains. However as the porphyria developed in the C57BL/6J strain at 5 weeks expression was induced more than 5 fold. So too was the induction of heme oxygenase 1 *(Hmox1)* implicating that the combination of UROD inhibition and increased HMOX1 activity was leading to a heme deficiency. All analyses considered were statistically significant with data from 4 or 5 mice. For particular genes of interest multiple clones were present on the arrays. Sometimes, although induction over controls was significantly greater in one strain than another, comparison of control mice demonstrated markedly different basal expressions that may be very pertinent to mechanisms of action.

Way forward

Both molecular mapping and genomics produce large amounts of data for analysis. At the present time we are in the process of trying to marry expression data with susceptibility loci. An important aspect of our candidate and unknown gene expressions is to determine their chromosomal positions to see if there are significant associations with quantitative loci. Besides strains with natural different susceptibilities, data from engineered strains such as *Cyp1a2* and *Ahr* null mice will be added into the comparisons. Eventually, we should be able to gain much greater insights into the molecular mechanisms of toxicity leading to pathology of dioxin in the liver as well as of other agents. Clarification of primary changes and secondary molecular pathological responses should be of assistance in short term studies of predictive toxicology.

References

Alizadeh AA, Elsen MB, Davis E, Ma C, Lossos IS, Rosenwald A, Boldrick JC, Sabet H, Tran T, Yu X, Powell JI, Yang Liming, Marti GE, Moore T, Hudson J Jr, Lu L, Lewis DB, Tibshirani R, Sherlock G, Chan WC, Greiner TC, Weisenburger DD, Armitage JO, Warnke R, Levy R, Wilson W, Grever MR, Byrd JC, Botstein D, Brown PO, Staudt LM (2000) Distinct types of diffuse large B-cell lymphoma identified by gene expression profiling. Nature 403:503-511

Fernandez-Salguero PM, Hilbert DM, Rudikoff S, Ward JM, Gonzalez FJ (1996) Aryl-hydrocarbon receptor-deficient mice are resistant to 2,3,7,8-tetrachlorodibenzo-*p*-dioxin-induced toxicity. Toxicol Appl Pharmacol 140:173-179

Fleming RE, Holden CC, Tamatsu S, Waheed A, Brunt EM, Britton RS, Bacon BR, Roopenian DC, Sly WS (2001) Mouse strain differences determine severity of iron accumulation in *Hfe* knockout model of hereditary hemochromatosis. PNAS 98:2707-2711

Greig AG, Francis JE, Kay SJE, Lovell DP, Smith AG (1984) Incomplete correlation of 2,3,7,8-tetrachlorodibenzo-*p*-dioxin hepatotoxicity with *Ah* phenotype in mice. Toxicol Appl Pharmacol 74:17-25

Gu Y-Z, Hogenesch JB, Bradfield CA (2000) The PAS superfamily: Sensors of environmental and developmental signals. Annu Rev Pharmacol Toxicol 40:519-561

Ikeda M, Ishii Y, Kato H, Akazawa D, Hatsumura M, Ishida T, Matsusue K, Yamada H, Oguri K (2000) Suppression of carbonic anhydrase 111 in rat liver by a dioxin-related toxic compound, coplanar polychlorinated biphenyl, 3,3',4,4',5-pentachlorobiphenyl. Arch Biochem Biophys 380:159-164

Meyer UA, Zanger UM (1997) Molecular mechanisms of genetic polymorphisms of drug metabolism. Annu Rev Pharmacol Toxicol 37:269-296

Nebert DW, Puga A, Vasiliou V (1993) Role of the Ah receptor and the dioxin-inducible [*Ah*} gene battery in toxicity, cancer, and signal transduction. Ann N Y Acad Sci 685:624-640

Oscarson M, Ingelman-Sundberg M, Daly AK, Nebert DW (2001) Human Cytochrome P450 (*CYP*) Alleles. http://www.imm.ki.se/CYPalleles/.

Shertzer HG, Nebert DW, Puga A, Ary M, Sonntag D, Dixon K, Robinson LJ, Cianciolo E, Dalton TP (1998) Dioxin causes a sustained oxidative stress response in the mouse. Biochem Biophys Res Commun 253:44-48

Smith AG, Clothier B, Robinson S, Scullion MJ, Carthew P, Edwards, R, Luo J, Lim CK (1998) Interaction between iron metabolism and 2,3,7,8-tetrachlorodibenzo-*p*-dioxin in mice with variants of the *Ahr* gene:A heptic oxidative mechanism. Mol Pharmacol 53:52-61

Smith AG, Clothier B, Carthew P, Childs NL, Sinclair PR, Nebert DW, Dalton TP (2001) Protection of the *Cyp1a2(-/-)* null mouse against uroporphyria and hepatic injury following exposure to 2,3,7,8-Tetrachlorodibenzo-*p*-dioxin. Toxicol Appl Pharmacol 173:89-98

Turton NJ, Judah, DJ, Riley J, Davies R, Lipson, D, Styles JA, Smith AG, Gant TW (2001) Gene expression and amplification in breast carcinoma cells with intrinsic and acquired doxorubicin resistance. Oncogene 20:1300-1306

Technical Development

(Genomics and Proteomics)

Application of gene arrays in toxicology

Matthew Bartosiewicz[1] and Alan Buckpitt[2]

[1] Department of Bioengineering, University of California San Diego, CA 92037
[2] Department of Molecular Biosciences, School of Veterinary Medicine, University of California Davis, CA 95616

Summary. The rapid development of new technologies capable of examining gene and/or protein expression on a global scale combined with the availability of either complete or nearly complete gene sequences from human and common rodent models has resulted in a paradigm shift in the way in which scientific questions are being addressed in many disciplines, including toxicology. The application of genomic tools to address important issues in toxicology include studies, which examine genome wide expression profiling in cells or tissues in response to toxicants as well as studies, focused on understanding the effects of sequence polymorphisms in the toxic response. The potential for using these approaches to understand mechanisms of action of toxicants and to assess the potential health effects of exposure to particular chemicals has received considerable attention and has been reviewed in several recent articles (Farr et al. 1999, Burchiel et al. 2001, Pennie et al. 2001, Fielden et al. 2001). This proceeding will discuss several issues, which we feel are important in the design and conduct of microarray studies along with applications of this in toxicology. The reader is referred to several recent articles for more information on the "how to" aspects of this technology (Bowtell 1999, Hedge et al. 2000).

Key words. In-Vivo, Toxicology, Toxicogenomics

Introduction

One of the primary goals in toxicity testing is to define the deleterious effects of new chemical compounds prior to their market release. The advent of new technologies and approaches, which improve the speed with which compounds can be screened for toxicity, are likely to result in considerable savings of money and time. Indeed, as the speed of toxicity testing increases, initial compound selection may be based not only on efficacy but on safety as well (Pennie et al 2000).

Toxicity studies currently rely on the use of animal bioassays to determine acute, subacute and subchronic toxicity, as well as carcinogenicity, and reproductive toxicity. The problems inherent in these tests, including cost, time and the sheer number of animals required, are recognized by both state and federal regulatory agencies such as the EPA, NIEHS, and the NTP (Cunningham 2000). Accordingly there has been a substantial effort to develop new methodologies, which are more time and cost effective.

Another recent technological advance, DNA microarrays, has been recognized for its potential use in toxicity testing (Lovett 2000, Medlin 1999, Rockett and Dix 1999). Briefly DNA arrays consist of a solid substrate (glass slides, nylon membranes) onto which fragments of genes (EST's) or oligonucleotides are attached at high density at predefined sites. Subsequent isolation of mRNA from treated and control (or diseased and normal) tissue/cells followed by labeling, hybridization, and reading can be used to assess the levels of expression of thousands of genes simultaneously. As suggested in several symposia, microarrays may provide a method to quickly categorize chemicals based on "fingerprints" of gene expression profiles thereby improving the efficiency of standard toxicity bioassays (Rockett and Dix 1999). Suggestions also have been made that this new technology may provide a high-throughput means to delineate mechanisms of action and predict the toxicity of unknown agents but much remains to be done to determine whether this is feasible (Pennie et al. 2001, Afshari et al. 1999, Nuwaysir et al. 1999). The fact that the technology is immature and that there are a number of platforms for printing arrays support the need for adequate validation before firm conclusions can be reached on the utility of microarray approaches in solving problems in toxicology (Yue et al. 2001). Additional work is needed to explore issues associated with the quantitative aspects of the technology and the parameters critical for toxicant prediction and mechanistic toxicology. The work described here has examined some of the early quantitative issues related to the use of this technology, and shows that gene expression profiles differ with various hepatotoxic agents and discusses some of our ongoing studies focused on using arrays to examine changes in the lung in response to exposure to multiple chemicals.

Predictive and mechanistic toxicology

Although the potential applications of microarrays in toxicology are extensive, there are several issues that require consideration when applying these technologies to mechanistic or predictive toxicology. The first of these issues is related to whether the toxicant's primary mode of action involves receptor interactions, which are transduced at the gene level, or whether there are other cellular targets (proteins, lipids), which begin the cascade of events leading to toxicity. For example, toxicants such as TCDD, which have as a primary mode of action changes in gene expression, are likely to lead to unique patterns of expression that are closely tied to the mechanisms of toxicity of the agent. Numerous other chemicals that act principally to disrupt membrane integrity, to alter mitochondrial ATP or to alter Ca^{2+} status in the cell would be expected to cause changes in gene expression as a downstream event resulting from disruption of homeostasis rather than being the triggering event which leads to toxicity. Nevertheless, it is still quite possible that the downstream events, which occur in response to toxicant exposure, will generate a fingerprint of specific responses of the cell to that toxicant. Perhaps the most important issue to be addressed by microarray studies is in the area of mechanistic toxicology where having additional information regarding the reaction of the cells to the injurant can provide a better understanding of a full range of events that occur in the cell in response to the toxicant. In turn, this allows investigators to generate well-defined hypotheses about the mechanism of action for a particular chemical, which can then be further analyzed with molecular and biochemical experiments.

In-vivo toxicology studies

Initially our laboratory constructed a DNA array that consisted of approximately 150 unique expressed sequence tags coding for DNA repair enzymes, heat shock proteins, and phase I and II metabolizing enzymes. This array layout was used to examine some of the important quantitative issues associated with this technique. Updated arrays containing 260 ESTs involved in phase I and phase II metabolism, heat shock, DNA repair, inflammation, transcription, and housekeeping were used in more recent work to examine issues related to the application of this technology to determine whether toxic chemicals would generate unique gene expression patterns. Many of the cDNAs for our arrays were chosen because they were known to be up or down regulated in response to the chemicals we wished to study. This provided a set of positive controls for each experiment and allowed comparison with the results in the published literature (Beyersmann and Hechtenburg 1997, Bhattacharjee et al. 1998, Dalton et al. 1999, Horn et al. 2000, Schiaffonati and Tiberio 1997). Dual labeling, Cy3 (control) and Cy5 (treated) experiments were conducted on these arrays. For analysis of the microarray results relative fluorescent units (RFU) were calculated using local median background subtraction (ArrayVision Software). The relative fluorescent units were then normalized to the median signal of probe for that slide. The change in gene expression for each spot was calculated as RFUtreated / RFUcontrol. For a more thorough explanation of the methods and results see references (Bartosiewicz et al. 2000, Bartosiewicz et al. 2001).

Quantitative issues associated with glass slide arrays

The goal of our initial work (Bartosiewicz et al. 2000) was to understand the potential limitations of microarray technology prior to developing arrays specifically for monitoring transcriptional changes in response to chemical exposure. The study was designed to investigate the dynamic range, sensitivity and experimental noise of the microarrays. Mice (4-5/group) were treated intraperitoneally (ip) with 10 different doses of β-naphthoflavone (0.15 mg/kg – 80 mg/kg) and gene expression was measured at a time point (8 hrs), which we had determined previously to result in the maximal induction of both cytochrome P450 1a1 and 1a2. Cyp1a1 and 1a2 expression was then monitored by both Northern blot and microarray as a function of β-NF dose. The results demonstrated that the sensitivity and dynamic range was slightly better with the microarrays than with the Northern blots. The variability introduced in DNA arrays by spotting and hybridization was considerably less than the interanimal variability. In general, the coefficient of variation in data between slides and between spots was less than 10% while variability of the response in the outbred mouse strain used in this work was generally on the order of 20-25%. Our early work also illustrated the need to account for differences in dye bias, which caused some of the genes being examined (metallothionein and alpha v integrin) to appear induced or repressed. To control for dye bias, separate controls were labeled with both Cy5 and Cy3 to establish a baseline expression level for all genes in all experiments. This baseline then allowed the significance of induction or repression of gene expression in response to β-NF to be determined.

Transcript profiling to determine whether different chemical classes provide unique "fingerprints"

Although several publications have suggested the use of gene expression arrays to categorize chemicals, there are only limited data supporting this approach (Buczynski et al. 2000). Initial, proof of principle studies determined whether unique gene expression profiles are generated in response to benzo(a)pyrene (BaP), $CdCl_2$ and trichloroethylene (TCE) as a function of dose in the liver (Bartosiewicz et al. 2001).

A more extensive study (Bartosiewicz et al. 2001) compared gene expression patterns in liver and kidney in response to 5 classes of chemicals (PAHs: benzo(a)pyrene (BaP), 3-methylcholanthrene (3-MC), DNA alkylators: dimethylnitrosamine (DMN), ethylnitrosourea (ENU), peroxisome proliferators: diethylhexylphthalate (DEHP), clofibrate, heavy metals: $CdCl_2$, $HgCl_2$ and oxidative stressors: CCl_4 bromobenzene) given intraperitoneally to mice. Time course experiments in mice were carried out in both kidney and liver for each chemical and dose response studies were used to further evaluate the transcriptional response of several of these chemicals. At each time point or dose 3-5 animals were used.

For each pair of chemicals tested, a similar pattern of gene expression was observed and this pattern was distinct from the other 4 classes of chemicals. In fact several of the genes induced by each class of compound were unique to those classes. Acyl-CoA thioesterase, Cyp4a10 and igf-binding protein 1 were only induced by the peroxisome proliferators, DEHP and clofibrate. Cyp1a1 was induced only by the PAHs, 3-MC and BaP. Monokine induced by gamma interferon and Bax Alpha were only induced by the DNA alkylators, DMN and ENU and MUP I and MUP II were only induced by the heavy metals, $CdCl_2$ and $HgCl_2$.

In most cases, the transcriptional response in the kidney was less than the response observed in the liver. The differential responses of liver and kidney are likely due to several factors. These include the fact that some of the toxicants in question have greater inherent toxicity in the liver, the toxicants reach higher concentrations in the liver in part because of the route of administration (intraperitoneal) or that the toxicants affect a higher percentage of the cell population. In this regard, the liver is comprised of a much more homogenous cell population than is the kidney. Toxicants such as CCl_4 which require metabolic activation by the cytochrome P450 monooxygenases will affect a small subset of the overall cell population in the kidney (the proximal tubular epithelial cells) and therefore, the magnitude of changes in gene expression occurring in this cell phenotype would be diluted considerably by mRNA obtained from all of the other cell types in the kidney. In general, the differential response of cells in heterogenous tissue is potentially one of the major stumbling blocks to the rapid and sensitive detection and classification of toxicants by gene expression analysis. Although advanced techniques such as laser capture microdissection can be used to select populations of target cells and can potentially solve some of the issues related to tissue heterogeneity, these techniques are time consuming and would not be useful in those cases where the target cells have not yet been identified.

Summary of in-vivo studies

In summary, we found the arrays to be as quantitative as standard methods that have generally been used to determine gene induction. A number of ambiguous results were eliminated by controlling for the dye bias of Cy3 and Cy5 as well as by conducting replicates for each dose and time point in our study. Provided that the experimental conditions are tightly controlled and that there are sufficient experimental replications robust results can be achieved with the microarrays. Additionally if some of the genes selected in the creation of a toxicological DNA array have been well studied with the particular toxicants in question, these genes can be useful as positive controls. This was particularly helpful in the studies discussed here, as the results were consistent with previous work that had focused on one or several genes.

For the eleven chemicals studied so far, each class has produced a similar pattern of gene induction unique from the others. From these data it appears that the microarrays will be useful for the classification of unknown compounds. This prediction may not always be sufficiently precise as to determine the exact class of chemicals to which a compound belongs (i.e. heavy metal) but, with further work, this technology may at least indicate a general grouping or mode of action for an unknown chemical. Several parameters were found to be particularly critical in obtaining meaningful data including 1) time, 2) dose, 3) tissue and 4) replicates. To accurately determine the potential hazards of unknown chemicals, by predicting a class or mode of action for that chemical requires that a full range of time points and doses be examined in several tissues. For example, clofibrate induced a number of genes at four hours but it was not until the eight-hour time point that those same genes showed significant induction by DEHP. Additionally the dose response study for DMN showed that significant induction was observed for Bax alpha at 10 mg/kg and 20 mg/kg but not at 25 mg/kg the dose used in the time comparison with ENU. Finally, assessing gene expression profiles in several tissues is also critical in that many chemicals target specific organs and meaningful alterations in gene expression are highly unlikely in non-target tissues. For all of the toxicants we tested, the liver and kidney were target organs for these compounds. However when gene expression was evaluated in the kidney after exposure to DMN (25 mg/kg) no significant induction was observed and it was only in the liver that this DNA alkylator altered the transcription of genes.

Current work

Our present work focuses on the use of transcript profiling to examine the markedly enhanced susceptibility of ozone exposed rats to the metabolically activated cytotoxicant, 1-nitronaphthalene. The importance of these studies stems from the fact that air in urban areas throughout the world is a complex mixture of oxidants, particulates and various organic chemicals. Studies conducted decades ago clearly demonstrated that multiple exposures to ozone resulted in lungs, which were subsequently tolerant to high exposures. In the process of determining whether ozone tolerance also conferred tolerance to metabolically activated chemicals, Paige and her coworkers (Paige et al. 2000) discovered that ozone exposed animals were substantially more susceptible to the pulmonary epithelial cell toxicity of 1-nitronaphthalene than are corresponding filtered air controls. Since 1-nitronaphthalene is a common copollutant with ozone in urban air these findings might have broad human health significance. We are applying microarray technology to begin to understand changes that occur in gene expression in ozone tolerant lungs, which would cause them to be more susceptible to 1-nitronaphthalene. To this end, we have prepared and sequenced cDNA libraries from both airways and parenchyma of both control and 1-nitronaphthalene treated rats. These have been utilized to construct "chips" containing approximately 9,000 unique elements (Shultz et al. 2001)). We intend to utilize these to define changes in gene expression during initial lung injury, repair and tolerance to ozone alone and in combination with 1-nitronaphthalene.

Acknowledgement

Work in the authors' laboratories has been supported by NIEHS 04699 and 09681.

References

Afshari CA, Nuwaysir EF, Barrett JC (1999) Application of complementary DNA microarray technology to carcinogen identification, toxicology, and drug safety evaluation. Cancer Res 59:4759-4760

Bartosiewicz M, Penn S, Buckpitt A (2001) Applications of gene arrays in environmental toxicology: Fingerprints of gene regulation associated with CdCl2, benzo(a)pyrene and trichloroethylene. Env Health Pesp 109:71-74

Bartosiewicz M, Trounstine M, Barker D, Johnston R, Buckpitt A (2000) Development of a toxicological gene array and quantitative assessment of this technology. Arch Biochem Biophys 376:66-73

Bartosiewicz MJ, Jenkins D, Penn S, Emery J, Buckpitt A (2001) Unique gene expression patterns in liver and kidney associated with exposure to chemical toxicants. J Pharm Exp Ther 297:895-905

Beyersmann D, Hechtenburg S (1997) Cadmium, gene regulation and cellular signaling in mammalian cells. Toxicol Appl Pharmacol 144:247-261

Bowtell DD (1999) Options available--from start to finish--for obtaining expression data by microarray. Nat Genet 21:25-32

Bhattacharjee A, Lappi VR, Rutherford MS, Schook LB (1998) Molecular dissection of dimethylnitrosamine (DMN)-induced hepatotoxicity by mRNA differential display. Toxicol Appl Pharmacol 150:186-195

Burchiel SW, Knall CM, Davis JW, Paules RS, Boggs SE, Afshari CA (2001) Analysis of genetic and epigenetic mechanisms of toxicity: potential roles of toxicogenomics and proteomics in toxicology. Toxicol Sci 59:193-195

Burczynski ME, McMillian M, Ciervo J, Li L, Parker JB, Dunn RT, Hicken S, Farr S, Johnson MD (2000) Toxicogenomics-based discrimination of toxic mechanism in hepg2 human hepatoma cells. Toxicol Sci 58:399-415

Cunningham MJ (2000) Genomics and proteomics: the new millennium of drug discovery and development. J Pharmacol Toxicol Methods 44:291-300.

Dalton TP, Shertzer HG, Puga A (1999) Regulation of gene expression by reactive oxygen. Annu Rev Pharmacol Toxicol 39:67-101

Farr S, Dunn RT (1999) Concise review: gene expression applied to toxicology. Toxicol Sci 50: 1-9

Fielden MR, Zachareweski TR (2001) Challenges and limitations of gene expression profiling in mechanistic and predictive toxicology. Toxicol Sci 60: 6-10

Hegde P, Qi R, Abernathy K, Gay C, Dharap S, Gaspard R, Hughes JE, Snesrud E, Lee N, Quackenbush J (2000) A concise guide to cDNA microarray analysis. Biotechniques 29: 548-560

Horn TL, Bhattacharjee A, Schook LB, Rutherford MS (2000) Altered hepatic mRNA expression of apoptotic genes during dimethylnitrosamine exposure. Toxicol Sci 57:240-249

Lovett RA (2000) Toxicologists brace for genomics revolution. Science 289: 536-537

Medlin J (1999) Timely toxicology. Env Health Persp 107:A256-A258

Nuwaysir EF, Bittner M, Trent J, Barrett JC, Afshari CA (1999) Microarrays and toxicology: The advent of toxicogenomics. Mol Carcinog 24:153-159

Paige RC, Wong V, Plopper CG (2000) Long-term exposure to ozone increases acute pulmonary centriacinar injury by 1-nitronaphthalene: II. Quantitative histo-pathology. J Pharmacol Exper Ther 295:942-950

Pennie WD, Woodyatt NJ, Aldridge TC, Orphanides G (2001) Application of genomics to the definition of the molecular basis for toxicity. Toxicology Letters 120:353-358

Pennie WD, Tugwood JD, Oliver GJ, Kimber I (2000) The principles and practice of toxicogenomics: applications and opportunities. Toxicol Sci 54:277-283

Rockett JC, Dix DJ (1999) Applications of DNA arrays to toxicology. Env Health Persp 107:681-685

Schiaffonati L, Tiberio L (1997) Gene expression in liver after toxic injury: analysis of heat shock response and oxidative stress-inducible genes. Liver 17:183-191

Shultz M, Jovanovich S, Gu E, Thomas R, Rank D, Penn S, Fanucchi M, Plopper C, Buckpitt A (2001) Development of a gene array for expression analysis in rat lung: Application to ozone and 1-nitronaphthalene. Toxicologist 60: 576 Abs

Yue H, Eastman SP, Wang BB, Minor J, Doctolero MH, Nuttall RL, Stack R, Becker JW, Montgomery JR, Vainer M, Johnston R (2001) An evaluation of the performance of cDNA microarrays for detecting changes in global mRNA expression. Nucleic Acids Research 29: E41-1

The ToxExpress™ Module *: applying gene expression data to predictive toxicology

Donna L. Mendrick

Vice President, Toxicology, Gene Logic Inc., 708 Quince Orchard Rd., Gaithersburg, MD 20878

Summary. One of the major challenges for the pharmaceutical industry is to identify the human safety of new chemical entities (NCEs) in a precise and high-throughput manner. NCEs with the lowest toxicity potential have the highest likelihood of success in preclinical and clinical development. Genome-wide gene expression studies targeted at understanding the impact a toxicant across highly diversified and multifunctional gene sets representing multiple pathways will revolutionize the way toxicologists evaluate human safety issues (1-6).

This article describes Gene Logic's ToxExpress™ Module, a system that helps evaluate NCEs for probable toxicity markers. The system combines three components: a proprietary reference database derived from toxicity studies on *in vivo* rats and primary rat and human cells, multiple gene expression analysis technologies, and powerful software solution tools.

The ToxExpress™ gene expression database is built with data generated from Affymetrix, Inc.'s GeneChip® microarrays (Rat Genome U34 and Human Genome U95 sets). In addition, the ToxExpress™ database incorporates data collected using Gene Logic's Restriction Enzyme Analysis of Differentially Expressed Sequences (READS™) technology. This open platform technology is designed to analyze expression values of known genes, novel genes, and Expressed Sequence Tags (ESTs).

Using proprietary Gene Logic bioinformatics and data mining tools as well as third party analysis software, researchers can mine the ToxExpress™ database to identify gene expression profiles, markers, and pathways that indicate toxicity patterns in specific target organs.

ToxExpress™ database information can help identify toxic verses non-toxic responses and help rank a compound's toxicity potential in early development phases. ToxExpress™ can also elucidate mechanisms of action involved with toxic response so that compound's development steps can avoid implicated pathways. In addition, gene expression data may be used to identify physiological changes before phenotypic events occur, helping toxicologists "see" potential biologically toxic effects before employing traditional toxicological testing methods.

Key Words. Toxicogenomics, Genomics, Gene expression, Microarrays

Introduction

The development of new technologies to rapidly screen combinatorial chemistry libraries has led to an increase in the number of NCEs in the drug development pipeline. This situation challenges drug safety scientists to implement higher-throughput toxicity assays. Traditional preclinical toxicology studies are time consuming, expensive, require high amounts of each NCE under study, and may not accurately predict human toxicity. The limited number of endpoints evaluated in preclinical toxicology (such as weight loss, activity changes, clinical chemistry, and histopathology) is estimated to be less than 100 in even the most comprehensive studies (2), indicating they may not adequately measure NCE human toxicity. A fast, high-throughput technology that can identify toxicity more precisely and using smaller quantities of NCE than traditional methods is needed. Analyzing changes in hundreds of thousands of genes has proven to be effective in addressing these needs.

A reference database of gene expression profiles for toxicity prediction

The ToxExpress™ Module contains gene expression profiles and associated data obtained from rats, *in vitro* rat primary cells, and *in vitro* human primary cells exposed to compounds that exert toxic effects in: 1. both humans and rats, 2. rats but not humans, and 3. humans but not rats. The use of these three sample types allows investigators to correlate gene expression changes between rats and humans and to help define *in vivo* and *in vitro* screens for proprietary drugs. Gene expression studies are performed with a candidate compound(s) and data is compared to the gene expression patterns of reference compounds in the ToxExpress™ Module. Genes showing abnormal expression responses to toxins may be effective markers for evaluating the toxicity of proprietary compounds.

All samples are studied using the Affymetrix, Inc. GeneChip® Human Genome U95 Set (A-E) and Rat Genome U34 Set (A, B, C). Approximately 60,000 human genes and ESTs are represented on the Human U95 set. The Rat Genome U34 Set, with ~24,000 known genes and ESTs, is used to monitor gene expression changes following exposure of rats and of primary rat cells to the same toxins as those utilized in the human studies.

The ToxExpress™ reference database includes expression data generated after treatment with multiple drug classes that have different pathologies and different target organs. Drug compounds are selected based on the type of toxicity within tissues (e.g. liver necrosis, steatosis, cholestasis, and non-genotoxic carcinogenesis) and the target organ for toxic damage. For liver studies, compounds selected are toxic only in the human, only in the rat, or in both species. The ToxExpress™ database also contains data on gene expression changes resulting from drugs that exert a heterogeneous response on general clinical and pathology parameters allowing investigators to fully leverage the gene expression data for accurate toxicity prediction. Currently Gene Logic has reference databases for liver and kidney toxicity. Reference databases for heart and other organ toxicity are under development.

Typical *in vivo* rat experiments consist of animals treated with vehicle and two doses of the test compound: a pharmacological dose and a toxic dose. Initial dose-range studies are conducted before running full gene expression experiments. Full gene expression experiments contain a minimum of 3 animals per group. Each experiment captures data at 3 time points that cover early and late expressed genes, allowing for statistical analysis.

Gene expression data empowers early development phase drug compound triaging. Currently, investigators can develop a list of differentially expressed genes. Quantitative differences in gene expression levels can be measured and potential drug compound toxicity levels may be predicted.

A percentage of the rat samples represented in the ToxExpress™ Module are also evaluated using Gene Logic's proprietary READS™ technology. This open platform, a patented differential display technology, examines changes in gene expression by selective PCR amplification and displays 3'-end restriction fragments of double-stranded cDNAs. This technology does not require prior gene identification. It measures differential expression (following toxin treatment) between studied genes (known novel and/or potentially new genes) and ESTs. Lower abundance genes can also be detected using READS™.

A software solution to data management

The ToxExpress™ database contains millions of data points encompassing gene, protein, enzyme, pathology, clinical chemistries, and urinalysis; all are linked to specific studies and individual samples. To manage this enormous amount of data, Gene Logic developed the GeneExpress® 2000 software.

The Gene Express® 2000 software can perform complex operations including: generating and comparing fold change results in gene expression levels between control and toxin-treated samples (Figure 1); examining where genes of interest lie in metabolic and regulatory pathways; and performing contrast analysis to detect statistically significant changes over a dose and time course. Gene expression results can also be exported into Gene Express® -accessible third party tools, including Microsoft Excel, S-plus, Spotfire®, GeneScreen™ and Partek Pro 2000™. In addition, ToxExpress™ data can be combined with information from various public databases including, GenBank (gene information on ncbi.nlm.nih.gov/ntrez.index.html), KEGG (enzyme information:genome.ad.jp/kegg/), and Swiss-Prot (protein information: expasy.ch.sprot.sprot-top.html).

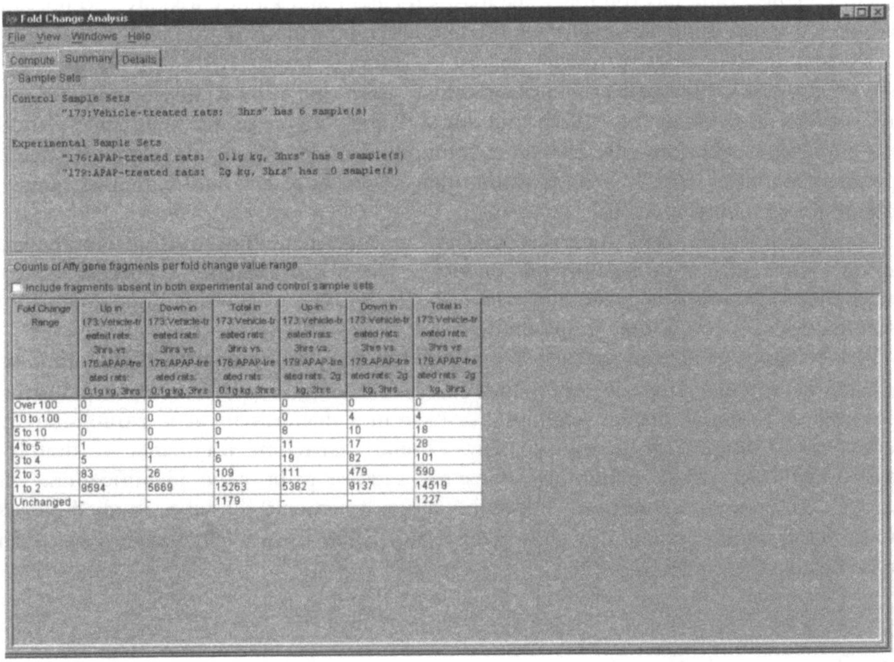

Figure 1. This fold change analysis compares vehicle treated rats (control) at 3 hours to acetaminophen (APAP) treated rats at 3 hours for low dose (0.1g/kg) and high dose (2 g/kg). Results in the table show the fold change range of gene expression data as compared to the control and the number of genes within each range from each study group. Individual genes in sample sets can also be examined.

The ToxExpress™ Module provides a comprehensive, versatile and accommodating platform that allows detailed gene expression analysis from one point of reference.

Conclusions

In summary, the ToxExpress™ Module is a powerful gene expression reference database built from studies of *in vivo* rat and *in vitro* primary rat and human cells treated with known non-proprietary compounds. Gene expression profiles from organs and sites predictive of classic toxicology responses provide the backbone of the database.

Gene expression data can be a powerful tool in triaging candidate drugs in the early development phase and in determining a compound mechanisms of action. GeneExpress® software streamlines data analysis and allows for data integration and analysis from outside sources while supporting multiple export formats and multiple third party analysis tools.

*** Patents are pending for various proprietary features incorporated into the ToxExpress ™ Module and related products, as well as proprietary data in the database.**
ToxExpress and READS are trademarks of Gene Logic Inc.
GeneExpress is a registered trademark of Gene Logic Inc.
GeneChip is a registered trademark of Affymetrix Inc.

References

1. Waring JF, Ulrich RG (2000) The impact of genomics-based technologies on drug safety evaluations. Annu Rev Pharmacol Toxicol 40:335-352
2. Farr S, Dunn RT, II (1999) Concise review: gene expression applied to toxicology. Toxicol Sci 50:1-9
3. Rockett JC, Dix DJ (1999) Application of DNA arrays to toxicology. Environ Health Perspect 107:681-685
4. Afshari CA, Nuwaysir EF, Barrett JC (1999) Application of complementary DNA microarray technology to carcinogen identification, toxicology, and drug safety evaluation. Cancer Res 59:4759-4760
5. Lovett RA (2000) Toxicologists brace for genomics revolution. Science 289:536-537
6. Pennie WD, Tugwood JD, Olliver GJA, Kimber I (2000) The principles and practice of toxicogenomics: applications and opportunities. Toxicological Sciences 54:277-283

Taming the mass of microarray data: systematic approaches to gene expression analysis

Anoop Grewal

Silicon Genetics, 2601 Spring St, Redwood City, CA 94063, U.S.A.

Summary. The volume of data generated from microarray experiments can be both promising and overwhelming for investigators. We show how software solutions for data analysis and data management can be used to make the data analysis process manageable, the workflow organized and the experimental results fruitful. GeneSpring™, Silicon Genetics' desktop solution for gene expression analysis is designed to analyze large gene expression data sets. The final objective is to generate a qualified list of candidate genes for more detailed study and to formulate additional hypotheses about pathways which mediate changes at the physiological level in response to the experimental conditions tested. We examine quality of raw data samples, set up normalization schemes, demonstrate filtering using statistical tools and exploit available gene annotations to mine the data further. Additionally, a companion tool, ScriptEditor™, is used to automate oft-repeated steps of the workflow to further streamline the data analysis process. Finally, we demonstrate the utility of having a shared repository for gene expression data within an institution through GeNet™, a web-enabled enterprise database solution for managing gene expression data.

Key words. Microarray, Workflow, Normalization, Quality control, Filtering, Data-mining, Database

Introduction

Molecular biologists have been quantitating RNA levels to analyze gene expression and gene function for over twenty-five years. But in this current age of high-throughput analysis and miniaturization, investigators can find themselves both compelled and confounded by the newfound capability to analyze thousands of genes simultaneously through the use of microarrays. By profiling the whole transcriptome in cells or tissue samples, the goal of identifying a set of genes whose expression levels change in response to altered experimental conditions is just the minimal expectation for the outcome of a study. GeneSpring, a desktop informatics solution, works hand-in-hand with technology for high-throughput gene expression data generation, the microarray. Users are enabled with sophisticated visualization options, statistical filters and data mining algorithms to allow users to prospect further for relationships and patterns, the identification of which should lead to new, better qualified hypotheses about mechanisms of action, gene function and dose-response relationships. A higher goal of toxicogenomics, as in the other genomics specialties, is to consolidate large volumes of microarray data into shared repositories, and further mine the resulting databases for insights that may only become apparent when such large collections of samples are available.

Table 1. Expression Data – Scales and Challenges

Example	Datapoints	Challenge
Clontech Atlas™ Toxicology II Array Genes per array: 588 Conditions: treated, untreated Replicates: no	~1200	-Normalize appropriately -Discard data with large contributions from noise -Identify genes with high fold-differences in expression
U.S. N.I.E.H.S. ToxChip V1.0 Genes per array: 2 100 Conditions: treated, untreated Replicates per condition: 2	~21 000	-As above -Check replicate quality -Identify genes with differential expression by statistical criteria
Affymetrix® Rat RGU34 GeneChip™ Genes per array: 26 000 Conditions: 0h, 1h, 2h, 4h, 24h Replicates per condition: 2	~260 000	-As above -Cluster genes
In-house Microarray Genes per array: 6 000 Conditions: 5 drug concentrations, 5 time points Replicates per condition: 6	~900 000	-As above -Examine data across multiple experimental variable dimensions
Core Facility Genes per array, average: 7 500 Arrays per day, average: 20	~250 000 per day	-As above -Manage multiple platforms -Automate repeated steps of the workflow
Institute's Database Genes per array, average: 7 500 Number of arrays per user, average: 10 Investigators: 20+	~1.5 M+	-As above -Find common data storage format for platforms -Archive all in-house data -Regulate access appropriately
Company/Consortium Databases Genes per array, average: 7 500 Conditions, average: 100s-1000s Replicates per condition: continuously increasing	1M – 1B and up	-As above -Enable cross-platform and cross-species analyses -Implement automated data-mining scalable to increasing data volumes

Each increase in the scale of the data presents unique challenges for handling and manipulating the data. At the first level, this starts with the choice of normalization schemes for a set of arrays that constitute a single experimental run and at higher levels this extends to the challenge of data mining gene expression results that span different platforms and organisms. Table 1 lists some of the specific tasks associated with each increase in data complexity and volume.

Workflow for an Experimental Run

Some investigators are alarmed to find that the complexity of the data analysis process exceeds that of the experimental design and data collection steps. Even with software tools that rapidly perform calculations over large, multi-dimensional sets, the user must make a series of informed choices suitable to his data type and platform. But while these individual choices and analytic decisions will vary with each researcher, a general workflow is common to most investigations. This workflow includes: Experimental Design, Sample Preparation, Data Collection, Data Import, Pre-processing, Filtering, Data Mining and Preparation of Figures & Tables and the latter five steps can all be performed with a single dedicated software solution such as GeneSpring™.

Experimental Design

Taking time at the earlier stages to design experiments that will most directly address the important questions will play a large role in the success of the subsequent analysis steps with regard to generating interpretable results, and generating well-qualified hypotheses based on the current investigation.

For instance, with a similar sample set, for example, compound A tested on mice at multiple times and doses, different investigators could very conceivably have different goals. One may wish to find the set of earliest genetic biomarkers responding to the drug class represented by the drug with the eventual goal of designing a diagnostic mini-array; another may be interested in comparing the collective gene profiles of toxic response to that for a better characterized agent of the same drug class; and a third researcher may be interested in identifying alternate pathways of action or further characterizing known pathways of action to include candidate new gene members from among the ESTs included in an array.

Assuming the number of arrays one can use is limited, a trade-off must be made between the number of different conditions tested and the number of replicates generated per condition. This choice has obvious statistical consequences. Figure 1 illustrates how the number of replicates used per condition can affect results.

Importing Data

Raw data files from most any format including custom formats can be imported as tab-delimited text files. GeneSpring will automatically recognize the format of common outputs (i.e. Axon Scanner files), then read, normalize and display the data based on default settings. In addition to information about gene name and signal intensity, data such as associated gene information, signal precision and data flags can be imported as well. Users will need to check and re-set default

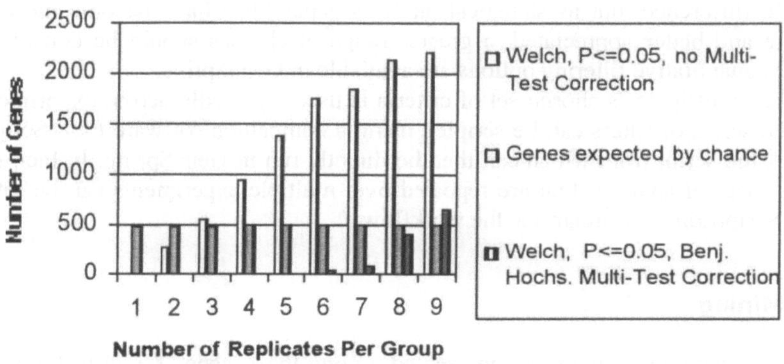

Fig. 1. Numbers of genes identified using statistical filters compared to number of replicates used in a sample study. Data was obtained from Welsh et al. (2001). For the two types of samples under study, normal human prostate and malignant human prostate, the number of genes obtained increased with the number of replicates used. The Welch t-test and Welch ANOVA were used for calculating p-values. *Abbreviations used*: P, p-value threshold; Multi-Test Correction, multiple testing correction; Benj. Hochs., Benjamini-Hochberg (Benjamin and Hochberg, 1995).

normalization settings as required and specify sample information for each sample (i.e. drug concentration, time, replicate number, etc). Before the actual data analysis begins, the user can additionally opt to (1) utilize a global error model, which estimates the error for each measurement as a function of the magnitude of its control or reference strength, (2) import gene-specific annotation from several public databases, and (3) set up multiple data modes and displays to facilitate interpretation of results.

Preprocessing Data

This step involves removing datapoints for a whole array or individual genes that are not sufficiently reliable to be worthy of further analysis. Excluding information about genes whose measurements may have large uncertainties avoids obscuring potentially interesting global trends that clustering or other data mining algorithms attempt to render apparent. Thus, removing such data predisposes the user to detecting potentially meaningful results.

Filtering

Once data has been processed, the first order of analysis is typically to identify genes which change significantly across the conditions tested. The criteria chosen to filter genes which exhibit a significant change between conditions can vary from investigator to investigator. Earlier this was often a minimum requirement of a 2-fold difference, but as statistical analysis capabilities have become more accessible and better appreciated, a greater range of choices should be considered. Various quantitative filtering options are available in GeneSpring.

If the investigator's chosen set of criteria is used repeatedly across experiments, then the series of filters can be scripted using a companion software tool, ScriptEditor™, the script from which can then be directly run in GeneSpring. In fact, most any series of analyses that are repeated over multiple experiments can be set up in the ScriptEditor to streamline the workflow.

Data Mining

How are functional categories represented among those genes found to be differentially expressed? Are recurring expression patterns shared by sets of genes? And if so, do these sets or clusters map to just one or a few functional categories? Can a prediction rule be set up that can be used to discriminate between samples of different classes (i.e. peroxisome proliferator agent-treated versus untreated)? Which of the genes identified as differentially expressed are ESTs for which functions have yet to be identified? Though the data mining step is often exploratory in nature, specific questions such as these can help the user to proceed through analysis in a focused manner.

A host of features are available in GeneSpring for data mining. Annotation tools integrate gene expression information with known biology. These include, the GeneSpider™ which will bring back standard annotations stored in GenBank by querying with GenBank accession numbers. Clustering tools, Complex-Correlations and Drawn-Genes-Correlations allow genes to be associated based on the similarity of their expression profiles. Principal components analysis is an algorithm to mine for the predominant expression profiles, while 'Predict Parameter Values' is a class prediction algorithm to help devise a diagnostic rule for predicting sample attributes (such as genotype, disease state, etc). And 'Find Potential Regulatory Sequences' is yet another tool designed to identify nucleotide sequences statistically over-represented in the upstream regions of a subset of genes grouped based on similarity of expression behavior.

Preparation of Tables & Figures

Once analyses are performed, the data can be exported in tabular format to generate tables for publishing. The same display options used earlier to visualize data

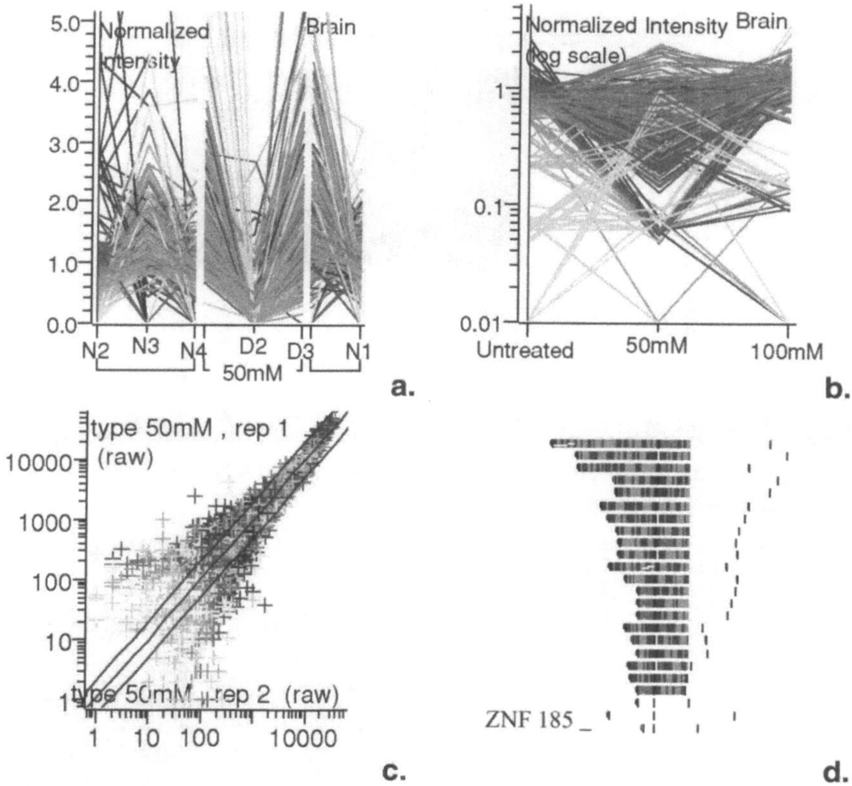

Fig. 2. Using Visualization Options to Enhance Interpretation of Results. Three of twelve view options available in GeneSpring are shown. Panels a and b represent the same data set in the 'Graph view' using different data modes and sample group choices. In (a), normalized intensity is displayed for each gene for array sample, while in (b), the logarithm of normalized intensities is taken and means from the set of replicates are plotted for each treatment. (c) A scatter-plot display allows a global quality control check for consistency between replicates. (d) The physical position view enabled data analysis efforts to locate differentially expressed genes specific to the known high penetrance chromosomal regions for prostate cancer for data from Welsh et al (2001).

meaningfully can now be used to generate figures for presentation. GeneSpring allows for saving results in the PICT image format enabling users to resize and manipulate elements of the resulting vector-based graphic to suit individual presentation requirements.

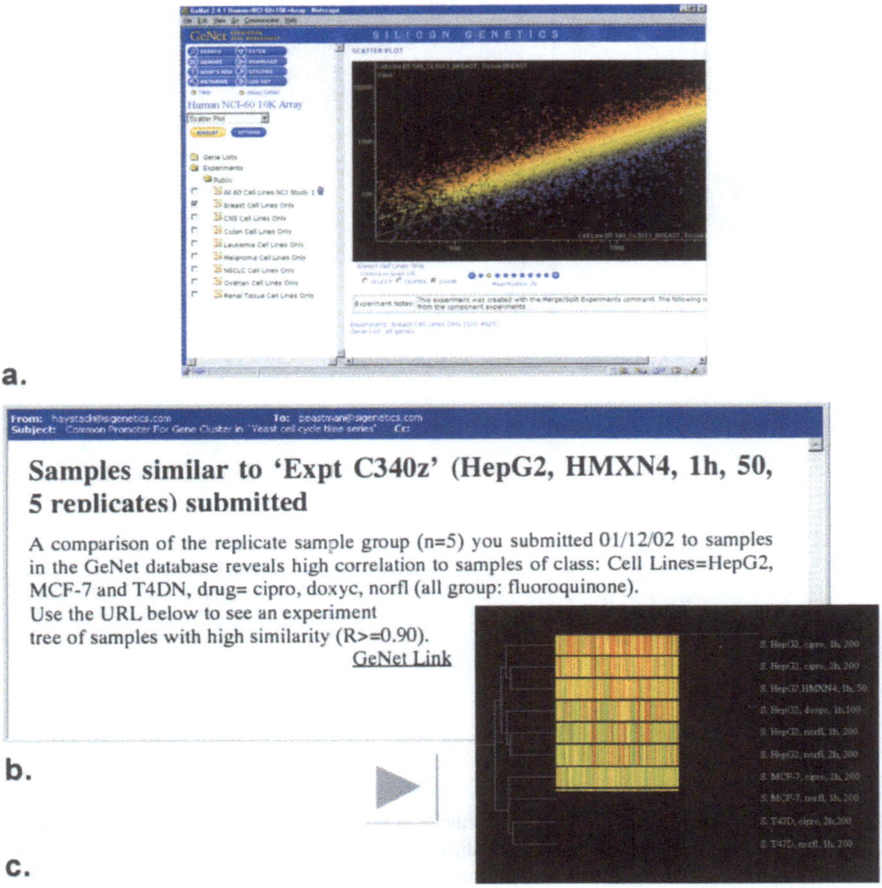

Fig. 3. GeNet and MetaMine. (a) A sample GeNet-enabled browser view of data accessed from a GeNet database. (b) A sample html-formatted email that could be received from a MetaMine search over a GeNet database.

Enterprise-Wide Data Management and Higher-Order Data Mining

The analysis process needn't end with individual data sets. Once added to a shared repository such as Silicon Genetics' GeNet™ database, the utility of the dataset is increased by its accessibility to investigators conducting similar studies and by contributing to a 'meta' data set which can be subjected to autonomous data mining such as that possible with the MetaMine™ product.

GeNet provides password-protected access to the database which can be a flat-file system or Oracle® database. Archived data can be accessed either through a web browser or directly through GeneSpring.

MetaMine is designed to discover valuable information among the incredibly large sets of gene expression data in a company or a research institution. MetaMine searches through the database of gene expression data and periodically reports interesting findings using email messages. MetaMine is designed to interface with GeNet, analyzing data generated by multiple GeneSpring users.

Conclusion

With the greater volume of data output, high-throughput gene expression analysis platforms have transformed the analysis process to one that requires software solutions specifically designed for gene expression. For microarray experiments, careful experimental design is as important as ever but a thorough, well-deliberated analysis for the generation of interpretable results is equally crucial. Software tools that not only perform computations over large volumes of data, but also enable the user to visualize the data in multiple ways and integrate biological information to facilitate interpretation add needed power to analysis. Data exploration can seem an unfocused meandering at first, but the workflow outlined above should not only help guide users through the process, but also lead them to make the most of otherwise under-utilized but useful features.

Finally, higher levels of data management require tools specifically customized for the specific volumes, tasks and workflow inherent to those larger datasets. Silicon Genetics has designed solutions for those endeavors, GeneSpring, Script-Editor, GeNet and MetaMine, and the four products in concert provide an integrated approach to managing all tiers of gene expression data management.

References

Benjamini Y, Hochberg, Y (1995) Controlling the false discovery rate: a practical and powerful approach to multiple testing. J Royal Stat Soc B 57:289-300

Welsh BJ, Sapinoso LM, Su AI, Kern SG, Wang-Rodriguez J, Moskaluk CA, Frierson HF Jr, Hampton GM (2001) Analysis of gene expression identifies candidate markers and pharmacological targets in prostate cancer. Cancer Res 61:5974-5978

Applications of ProteinChip® Systems in Toxicological Assessment

Lee O. Lomas

Ciphergen Biosystems Inc., 6611 Dumbarton Circle, Fremont, CA 94555, USA

Summary. Pharmaceutical companies are continually under pressure to develop cost-effective and safe treatments for an increasing number and complexity of medical conditions. As the complexity of drug design and function increases, it becomes clear that more rigorous methods of testing drug candidates for safety compliance is paramount. As a result, pharmaceutical companies have turned to the disciplines of toxicogenomics and toxicoproteomics in the hopes of developing predictive models for toxicological assessment. Although gene-based analysis are extremely powerful, they do not provide an in-depth or global picture of the spatial and temporal patterns of protein expression nor do they reveal the extent to which proteins are post-translationally modified. ProteinChip® Array-based systems are now emerging as a complementary approach to gene array technologies in the assessment of drug toxicity. These systems utilize chemically or biochemically activated surfaces to directly capture proteins based on their chemical/biochemical nature from biological samples, with direct detection by mass spectrometry. This technology therefore provides a high-throughput method for developing predictive models at the proteomic level to assess drug toxicity.

Key words. ProteinChip Arrays, proteomics, toxicology, mass spectrometry

As the complexity of drug design increases, more rigorous methods to assess drug toxicity must be established. Development of new biochemical methodologies analogous to the established endpoint assays (eg aldolase, aminotransferase, dehydrogenase, creatinine kinase, and urine creatinine) can be difficult, time consuming and expensive. Additionally, many of these markers lack the specificity of identifying the affected organ (Timbrell JA, 1998). With the completion of a number of genomic sequences, including human, functional and comparative genomics has become extremely useful tool used to test the response of cells or organisms to drug candidates varying experimental conditions. Gene-based microarray technology has enabled researchers to survey a large number of genes transcripts in order to determine and compare relative mRNA expression levels before and after drug challenge. In many situations, however, DNA/mRNA-based analysis is not able to provide an in-depth or global picture of the spatial and temporal patterns of protein expression nor do they reveal the extent to which proteins are post-translationally modified. Relying solely on genomic analysis to assess toxicological effect could allow an unsafe drug candidate to progress to the final stages of clinical trial, or worse be introduced into the public, before the toxic effect is discovered.

Toxicoproteomics offers the possibility to associate differential expression profiles with toxicological effect at the protein level. At present this approach has primarily involved the use of 2-D gel electrophoresis (2D-PAGE), however, this technique suffers from a number of technical limitations. 2D-PAGE is time-consuming, labor-intensive, requires significant technical expertise and lacks reproducibility (Jenkins RE and Pennington SR, 2001). As an alternative to 2D-PAGE, protein biochips are emerging as a simple-to-use high-throughput technology that offers the capability of differential expression analysis at the protein level in response to drug challenge.

Currently, only one protein biochip system is commercially available. This system utilizes patented surface enhanced laser desorption/ionization (SELDI™) technology combined with time-of-flight mass spectrometry (TOF-MS; Weinberger S, Morris TS, Pawlak M, 2000). Using this ProteinChip® System, subsets of proteins can be selectively captured on the array surface directly from a biological sample in a single step. Selectivity for the proteins captured is based on either biochemical characteristics (eg charge, hydrophobicity, metal affinity) or biologically defined by covalently attaching a target protein (eg antibody) to the array surface for the capture of a specific protein (eg antigen; figure 1; Merchant M and Weinberger SR 2000).

During the discovery phase to reveal candidate markers of toxicity, multiple ProteinChip Array surfaces and washing conditions can be empirically explored in parallel with a limited sample size to effectively resolve hundreds of proteins at a time. The output from such an experiment then provides a qualitative and relative quantitative representation of the peptides and proteins present in the biological sample. By comparing such profiles generated before and after drug treatment, it is possible to identify candidate biomarkers indicative of treatment (figure 2). Once a candidate marker of toxicity is discovered, validation becomes essential to demonstrate a correlative effect between the candidate biomarker and drug treatment. Using the same type of ProteinChip Array (eg anionic, cationic, hydrophobic, or metal affinity) and washing conditions used to discover the marker, it now becomes possible to process hundreds of samples in a few days for full validation. Final identification of a biomarker comes only after full validation. Identification of the marker is typically based on the enzymic digestion of the marker followed by peptide mass fingerprinting, or additionally, peptide sequencing by MS. Although peptide mass fingerprinting can be done effectively using the ProteinChip System mass reader, peptide

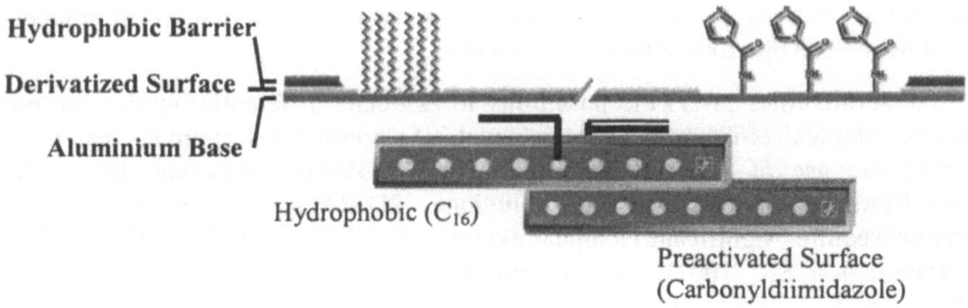

Figure 1. ProteinChip® Arrays are comprised of chemically or biochemically activated targets or spots. The interactions of chemically activated spots are based on standard chromatographical principles and include hydrophobic, ionic (both anionic and cationic), immobilized metal affinity capture (IMAC) and silica oxide interactions. Biochemically activated spots provide the chemistries for covalent coupling of a target (eg antibody) to the surface and are used for subsequent capture of an antigen from a complex biological sample.

sequencing typically requires a MS/MS instrument capable of collision-induced dissociation (CID). Peptide sequencing by mass spectrometry directly from the ProteinChip Array is now realized by the development of a ProteinChip Interface that can be accommodated by either the QSTAR™ (Applied Biosystems) or Q-Tof™ (Micromass) instruments.

One of the first examples demonstrating the suitability of SELDI-based ProteinChip array technology was used in the identification of a predictive toxic marker in 2,3,5,6 tetramethyl-*p*-phenylenediamine (TMPD) treatment in a rat model system (Dare TO, manuscript in preparation). Using this model, a single marker indicative of TMPD toxicity was detected from urine samples, validated and subsequently identified by peptide mass fingerprinting as parvalbumin-α. This marker was found to correlate with both histological indications of muscle toxicity and serological endpoint analysis using a combination of alanine- and aspartate-aminotransferase activity.

The use of single biomarker as a predictor of toxicity in every case is, however, unlikely because of the complexity of interactions at the cellular level and also the often subtlety of the toxic effect. Multiple-marker detection is, therefore becoming a common theme in proteome-based toxicological

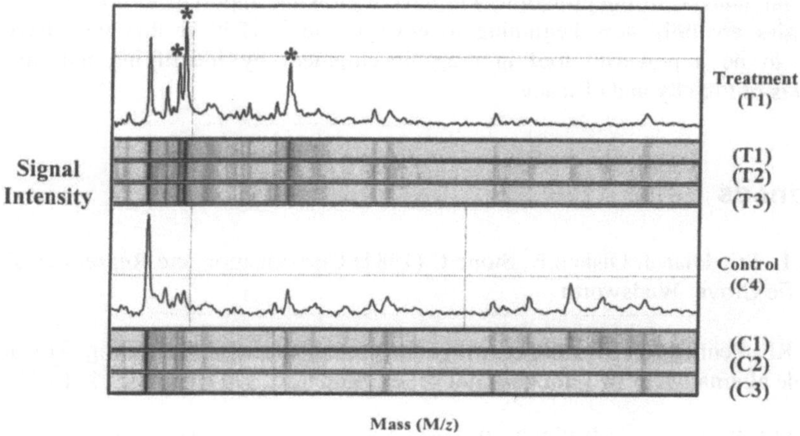

Figure 2. The use of ProteinChip Arrays to detect differentially regulated protein species. After treatment with a drug therapy known to cause tissue toxicity, tissue extracts were prepared and profiled on a series of chromatography-based ProteinChip Arrays. Under specific conditions, several proteins were discovered to be up-regulated in response to drug treatment (stars). Present are data from three treated and three control animals. Both 'spectral view' and theoretical density map or 'gel' view are displayed.

applications. To address these issues, powerful multivariate data analysis tools have been employed to mine the data effectively for the identification of predictive protein patterns. One such tool specifically adapted for the ProteinChip System is Biomarker Patterns™ Software, based on the CART algorithm technology (Breiman L et al 1984). Biomarker Patterns Software represents a robust decision-tree based tool for data mining, predictive modeling, and data preprocessing. Biomarker Patterns automatically searches for important patterns and relationships, uncovering hidden structure even in highly complex data. These correlative associations can then be used to generate accurate and reliable predictive models for applications in predictive diagnostics and toxicology (Fung E and Enderwick C, 2002).

There is little argument that the future challenges associated with safety assessment of new drug candidates will continue to become more complex and costly. With the considerable pressures on pharmaceutical companies to continually produce new, more effective drugs at reasonable cost, it is clear that effective drug management is critical and technologies that quickly identify drug candidates displaying toxicity during the initial phases or trials will be adopted quickly. Genomic technologies techniques have become powerful tools to address many of these concerns, however, it is clear that analysis of the proteome will have a growing importance. Protein biochip technologies are only now beginning to emerge and it is clear that they have the potential to be a powerful tool in drug development by identifying non-invasive biomarkers of toxicity and efficacy.

References

Breiman L, Friedman J, Olshen R, Stone C (1984) Classification and Regression Trees. Pacific Grove: Wadsworth

Jenkins RE, Pennington SR (2001) Arrays for protein expression profiling: Towards a viable alternative to two-dimensional gel electrophoresis? Proteomics 1: 13-29.

Merchant M, Weinberger SR (2000) Recent advancement sin surface-enhanced laser desorption/ionization-time of flight-mass spectrometry. Electrophoresis 21: 1164-1167.

Timbrell, J.A. (1998) Biomarkers in Toxicology 129: 1-12

Weinberger S, Morris TS, Pawlak, M (2000) Recent trends in protein biochip technology. Pharmacogenomics 1:395-416.

Experimental Applications (1)

Oncology Research

Toxicogenomics: making progress by maintaining a focus on the fundamentals of toxicology

Jay I. Goodman

Department of Pharmacology and Toxicology, Michigan State University
B-440 Life Sciences Building, East Lansing, Michigan 48824 U.S.A.
Email: goodman3@pilot.msu.edu

Summary. The study of gene expression is based upon the implicit assumption that altered expression is a key component underlying toxicological outcomes. A goal is to discover patterns of expression that can provide insight regarding genes that are involved in particular types of toxicity. This can enhance our understanding of key events that contribute to toxicity and there is considerable interest in identifying gene expression patterns that are indicative of toxicity, prior to its detection by classical means. The application of genomic technology to toxicology provides exciting new opportunities. However, toxicogenomics is a tool, it will not solve all of our problems. Key, basic aspects of risk assessment are: 1) rational selection of doses to be tested; 2) dose-response relationships, including the existence of thresholds; 3) species to species extrapolation issues; and 4) exposure assessment. These should be taken into account when genomic technologies are employed. In particular, one needs to be cognizant of the fact that dose influences mechanism, mechanism changes with dose and, therefore, effects that are produced by high doses do not necessarily occur at low doses. Furthermore, a change in the pattern of gene expression is not necessarily indicative of toxicity. In order to succeed as we embrace new molecular tools and incorporate them into research in toxicology, including risk assessment, it is imperative that the fundamental principles of toxicology are kept at the forefront.

Key words. epigenetics, genomics, metabolome, methylation, proteome, risk assessment, toxicology, toxicogenomics, transcriptome

Introduction

Toxicologists are embracing the developing field of genomics, the study of DNA sequence/organization and the relationship of this to function at the cell, tissue and whole plant/animal level (Farr and Dunn, 1999; Rodi et al., 1999; Lovett, 2000; Pennie et al., 2000; Smith, 2001; and Tennant, 2002). Genomics encompasses the study of the transcriptome (mRNA synthesis, gene expression patterns), proteome (expression of proteins), and metabolome (metabolite patterns). Basically, we can think of the genome (DNA sequence) as indicating what could happen, the transcriptome as what would happen and the proteome/metabolome as what is happening. Furthermore, genomics also involves a consideration of the methylome (DNA methylation status, an epigenetic process that plays a key role in the regulation of gene expression). As increased resources are directed towards identification of genotypes that affect susceptibility to toxicity it is important to understand that a strategy oriented solely towards detection of mutations/polymorphisms may be too limited. Epigenetic regulation/misregulation of genes is likely to be a key determinant of toxicity (Counts and Goodman, 1995; Robertson and Jones, 2000; Robertson and Wolffe, 2000; and Watson and Goodman, 2002). Indeed, toxicologists are positioned uniquely to contribute to the development of genomics, in addition to employing the tools that others bring forward. The focus of this manuscript is on the current interest in the study of gene expression, i.e. mRNA synthesis, the transcriptome. This includes the application of new, developing technology platforms to basic research aimed at discerning mechanisms of toxicity and to toxicity testing.

Gene expression arrays - toxicogenomics

Within this context, the study of gene expression is based upon the implicit assumption that (with the possible exception of necrosis that occurs rapidly) altered expression is a key component underlying toxicological outcomes. A goal is to discover patterns of expression that can provide insight regarding genes that are involved in particular types of toxicity. This can enhance our understanding of key events that contribute to toxicity. Additionally, there is considerable interest in identifying gene expression patterns that are indicative of toxicity, prior to its detection by classical means (e.g., histopathology and/or altered organ function) that can be more labor- and time-consuming. The idea here is that if a causal relationship can be demonstrated between early alterations in gene expression and frank toxicity, assessing the former could decrease the requirement to monitor the later (Farr and Dunn, 1999; Pennie et al., 2000). DNA chips, or microarrays, allow quantitative comparisons to be made of the expression levels of potentially many thousands of genes between different biological samples. In general, there are two approaches that may be taken to employ microarrays. Inductive logic involves the formation of rules from particular observations. The use of inductive arrays involves the measurement of thousands of genes, including expressed sequence tags (ESTs), and the "mining" of the resulting data for patterns that are associated statistically with the biological outcome. While we may not know the function of a given set of genes/ESTs, the altered pattern of expression could provide a signal for a particular outcome.

Deductive logic begins with the general rule and proceeds to the particular. Using custom designed arrays to profile a limited number of genes whose function is understood permits one to profile the activity of a chemical by the functions it elicits (Dunn and Farr, 1999). Indeed, toxicology is one of the biomedical sciences that are very well positioned to take advantage of the recent molecular biotechnology advances.

Making progress by focusing upon basic principles of toxicology

Toxicogenomics is a tool that if employed properly can be very helpful with regard to advancing toxicological sciences. However, It is important to understand that toxicogenomics does not represent a means to solve all of our problems. Legitimate concerns have been raised regarding the potential for misinterpretation of genomic analysis particularly in the context of safety assessment (Pennie, et al., 2000). For example, a change in the pattern of gene expression is not necessarily indicative of toxicity. These data need to be "anchored" to known toxic responses *in vivo*. After all, we do not want to employ new approaches in order to obtain more wrong answers faster! The application of genomic technology to toxicology provides exciting new opportunities. However, these opportunities will prove fruitless unless toxicologists recognize that new technologies impose a greater demand to exercise expert judgment on the meaning of the data produced and to apply common sense when balancing risks and benefits (Smith, 2001). To a large extent, safety evaluation of chemicals involves administration of relatively high doses of the compound in question to a small number of animals to characterize the effect(s), followed by an extrapolation to a large number of people who might be exposed to a relatively low dose of the compound. Key, basic aspects of risk assessment are (Conolly, et al., 1999): 1) rational selection of doses to be tested, and currently there is too much of a tendency to employ doses that are excessively high; 2) dose-response relationships, including the existence of thresholds; 3) species to species extrapolation issues; and 4) exposure assessment. These should be taken into account when genomic technologies are employed. In particular, one needs to be cognizant of the basic tenet of toxicology: "The dose makes the poison." Dose influences mechanism, mechanism changes with dose and, therefore, effects that are produced by high doses do not necessarily occur at low doses (Goodman, 1998). Furthermore, when using gene expression data to characterize and compare the effects of different compounds one must understand that the pattern of expression is expected to be different depending upon where one is on the dose/time response curve, e.g., the altered pattern of expression elicited by a chemical is not expected to be the same at the low vs. high end of the dose-response curve. Therefore, in order to compare different chemicals, their ability to affect gene expression must be evaluated under experimental conditions where the doses of each are at similar points on their respective dose-response curves for a particular toxic end point. Additionally, it is important to obtain a firm understanding of the variability in gene expression within the control population prior to evaluating the effect of a particular compound (Hughes et al., 2000; Pritchard, et al., 2001). While inclusion of these controls is expensive, making decisions based upon wrong information can be much more costly. These are exciting times to be involved in science, particularly toxicological sciences. The genomics revolution is aiding in advancing the biological sciences in an astounding fashion. In order to succeed as we embrace new molecular tools and incorporate them into research in toxicology, including risk assessment, it is imperative that the fundamental principles of toxicology are kept at the forefront.

References

Afshari CA, Nuwaysir EF and Barrett JC (1999) Application of complementary DNA microarray technology to carcinogen identification, toxicology and drug safety evaluation. Cancer Res. 59: 4759-4760.

Conolly RB, Beck BD and Goodman, JI (1999) Stimulating research to improve the scientific basis of risk assessment. Toxicol. Sci. 49: 1-4.

Counts JL and Goodman JI (1995) Alterations in DNA methylation may play a variety of roles in carcinogenesis. Cell 83:13-15.

Farr S and Dunn RT, III (1999) Concise Review: Gene expression applied to toxicology. Toxicol. Sci. 50:1-9.

Goodman JI (1998) The traditional toxicologic paradigm is correct: Dose influences mechanism. Env. Hlth. Perspect. 106: 285-288.

Hughes TR, Marton MJ, Jones AR, Roberts CJ, Stoughton R, Armour CD, Bennett HA, Coffey E, Dai H, He YD, Kidd MJ, King AM, Meyer MR, Slade D, Lum PY, Stepaniants SB, Shoemaker DD, Gachotte D, Chakraburtty K, Simon S, Bard M and Friend SH (2000) Functional discovery via a compendium of expression profiles. Cell 102: 109-126.

Lovett RA (2000) Toxicologists brace for genomics revolution. Science 289: 536-537.

Pennie WD, Tugwood JD, Oliver GJA and Kimber I (2000) The principles and practice of Toxicogenomics: Applications and opportunities. Toxicol. Sci. 54: 277-283.

Pritchard CC, Hsu L, Delrow J and Nelson PS (2001) Project normal: Defining normal variance in mouse gene expression. Proc. Nat. Acad. Sci. USA 98: 13266-13271.

Robertson KD and Jones PA (2000). DNA methylation: past, present and future directions. Carcinogenesis 21: 461-467.

Robertson KD and Wolffe AP (2000). DNA methylation in health and disease. *Nature Reviews Genetics* **1**, 11-19.

Rodi CP, Bunch RT, Curtiss SW, Kier LD, Cabonce MA, DaVila JC, Mitchell MD, Alden CL and Morris DL (1999) Revolution through genomics in investigative and discovery toxicology. Toxicologic Pathol. 27: 107-110.
Smith LL (2001) Key challenges for toxicologists in the 21st century. Trends in Pharmacol. Sci. 22: 281-285.
Tennant RW (2002) The National Center for Toxicogenomics: Using new technologies to inform mechanistic toxicology. Env. Hlth. Perspect. 110: A8-A10.
Watson RE and Goodman JI (2002) Epigenetics and DNA methylation come of age in toxicology. Toxicol. Sci., in press.

A microarray analysis of differential gene expression associated with the development of doxorubicin resistance in breast carcinoma

Timothy W. Gant, Nicola J. Turton, Joan Riley, David J. Judah, Reginald Davies, Doron Lipson, Jerry A. Styles, and Andrew G. Smith

MRC Toxicology Unit, University of Leicester, Leicester. LE1 9HN UK

Summary. One of the major complications of cancer chemotherapy is the development of a multidrug resistance (MDR) phenotype. Genes known to be major contributors to this phenotype include the ABC transporter family, DNA repair class genes and phase II conjugators such as the GST family. However, though there may be predominant contribution to the MDR phenotype from one gene or family the overall phenotype probably results from the contribution of a cohort of genes, both of downstream effecter and upstream regulatory classes. Additionally, other genes differentially expressed in the MDR verses drug sensitive states may result in other phenotypic traits that may also complicate treatment, for example an increased propensity towards invasiveness. We have applied microarray analysis to breast carcinoma cells with intrinsic and acquired doxorubicin resistant phenotypes and compared the gene expression profiles obtained with each other, and with those gathered from various other drug sensitive breast carcinoma cell lines (Turton et al. 2001). The analysis has been applied at both the mRNA and genomic levels. Some expected changes in gene expression (mRNA analysis) and amplification (genomic DNA analysis) such as that for the multidrug resistance 1 (*ABCB1*) gene were detected. However, some unexpected associations of differential gene expression were found, some associated with known phenotypic traits, and some that may offer alternative targets for chemotherapy. The similarity of gene expression profiles between the cells with intrinsic and acquired doxorubicin resistance indicated that a tumor with an acquired doxorubicin resistant phenotype arises as a result of drug driven selection of cells with an intrinsic resistant phenotype from the majority drug sensitive population.

Keywords. Breast carcinoma, doxorubicin resistance, microarray analysis, multidrug resistant

Development of a drug resistant phenotype

Intrinsic and acquired drug resistant phenotype

Intrinsic resistance is a term used to describe the phenomenon whereby a cell or group of cells has a pre-existing drug resistance. In contrast acquired resistance in a tumor describes the process by which the tumor develops drug resistance during the course of drug exposure (Hayes and Wolf 1990). The development of an acquired resistance to cancer chemotherapy in a tumor has been hypothesized to arise from the evolutionary selection of a cell with an intrinsic resistance under the pressure of chemotherapy (Fig. 1). As such the acquired resistance developed by the tumor would be a result of clonal selection of the one cell that had an intrinsic resistance, and would therefore be permanent. There is though also the possibility that resistance can develop through the differential expression of genes required to confer the resistance as a direct result of drug exposure. Such a resistance would not be permanent as induction of the required gene expressions would be dependent on the continued drug exposure. The likelihood of finding a cell with a permanent intrinsic resistance within a population of otherwise sensitive cells, for any given mutation rate, has been estimated by Goldie and Coldman (Goldie et al. 1982), who also articulated that case for the use of alternating non-cross resistant chemotherapy to mitigate the effects of developing drug resistance.

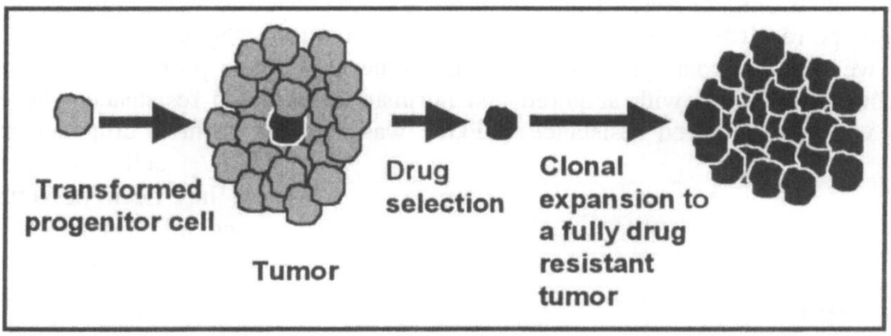

Fig 1. Acquisition of an MDR phenotype in tumors. A tumor can gain an acquired drug resistant phenotype as a result of drug selection of cells with an intrinsic drug resistance phenotype. These cells form part of the population and have an intrinsic resistant phenotype. This higher the mutation rate of the tumor the more likely is the existence of these cells. On exposure to drug they are given a growth advantage as a result of the selective killing of the drug sensitive cells and grow through to form a tumor with an acquired drug resistant phenotype.

However, the strategy of alternating non-cross resistant chemotherapy fails when there is the development of a multidrug resistant (MDR) phenotype in response to exposure to a single chemotherapeutic agent. Such a phenotype can be produced by the over-expression of a single gene the Multidrug resistance 1 (*ABCB1*) gene. This gene is able to confer a MDR phenotype because it operates as a broad specificity membrane pump, utilizing ATP to pump chemotherapeutic drugs from the cells against their concentration gradient (Endicott and Ling 1989). Thus cells which are overexpressing this gene will have resistance not only to the original drug molecule but also to all others which are substrates for the pump (Harrison 1995). The phenomenon of the development of resistance to other drugs after exposure to only one is called MDR.

Some tumors are intrinsically MDR as a result of *ABCB1* gene expression in the absence of drug exposure. This usually occurs in these tumor types because they have arisen from cells that *per se* express the *ABCB1* gene, such as those of the colon and the liver (Weinstein et al. 1991). However, for other tumors type a MDR phenotype develops as a result of chemotherapy, often as a result of *ABCB1* over-expression (Weinstein et al. 1991). One of the most common malignancies in this category is breast carcinoma.

Microarray analysis to explore the similarity of cells with intrinsic and acquired MDR phenotypes

In this study we set out to ask the question does an acquired MDR phenotype in breast carcinoma arise via the selection of a small number of cells with an intrinsic resistant phenotype by selection from the majority drug sensitive population as a result of chemotherapy? The converse question was, does a tumor MDR phenotype result from transient drug induced expression of specific genes? To do this we used microarrays to compare the gene expression profiles from breast carcinoma cell lines with acquired and intrinsic doxorubicin resistance. The cell line with the acquired resistance (HPGP) was selected from a drug sensitive MCF7 cell population and has been previously described (Davies et al. 1996). The intrinsically resistant cell line was selected from a drug sensitive population of MDA468 cells using the MRK16 antibody (Tsuro et al. 1989). MRK16 binds the extra-cellular loops of the *ABCB1* gene product P-glycoprotein, and so can be used to identify cells with this

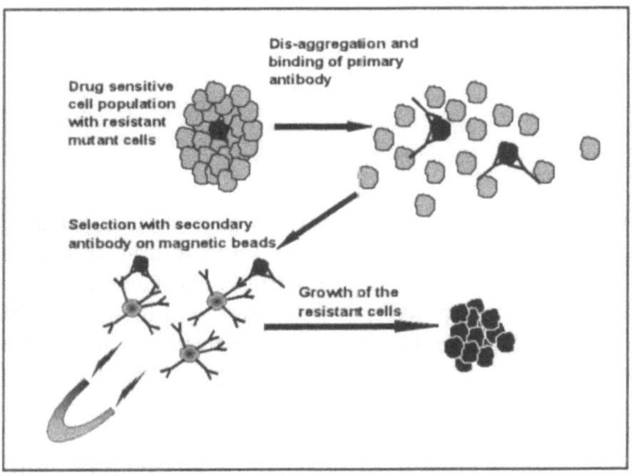

Fig 2. Selection of cells with an intrinsic resistance phenotype. These cells were selected from the drug sensitive population by the use of the antibody MRK16 that binds to the extracellular loops of the *ABCB1* gene product P-glycoprotein. After binding of the antibody to the few cells in the population that had an intrinsic resistant phenotype they were separated from the majority population by the use of magnetic beads with to which the secondary antibody was attached. They were then allowed to grow on to form an intrinsically drug resistant clone of the original drug sensitive population.

protein in without compromising the integrity of the cell membrane (Vasudevan et al. 1998). Using a secondary antibody bound to magnetic beads it was then possible to purify a small number of cells, designated MDA16, which have an intrinsic resistance through the expression of *ABCB1* (Turton et al. 2001) (Fig 2).

Using microarrays prepared in our laboratory (http://www.le.ac.uk/cmht/twg1/array-fp.html) we have determined the patterns of gene expression in the acquired and intrinsically resistant cell lines relative to their drug sensitive progenitors, and also in relation to other breast carcinoma cell lines (Turton et al. 2001). A clear gene expression pattern related to the multidrug resistant phenotype was obtained which was common to both the HPGP (acquired) and MDA16 (intrinsic) multidrug resistant cell lines (Fig 3).

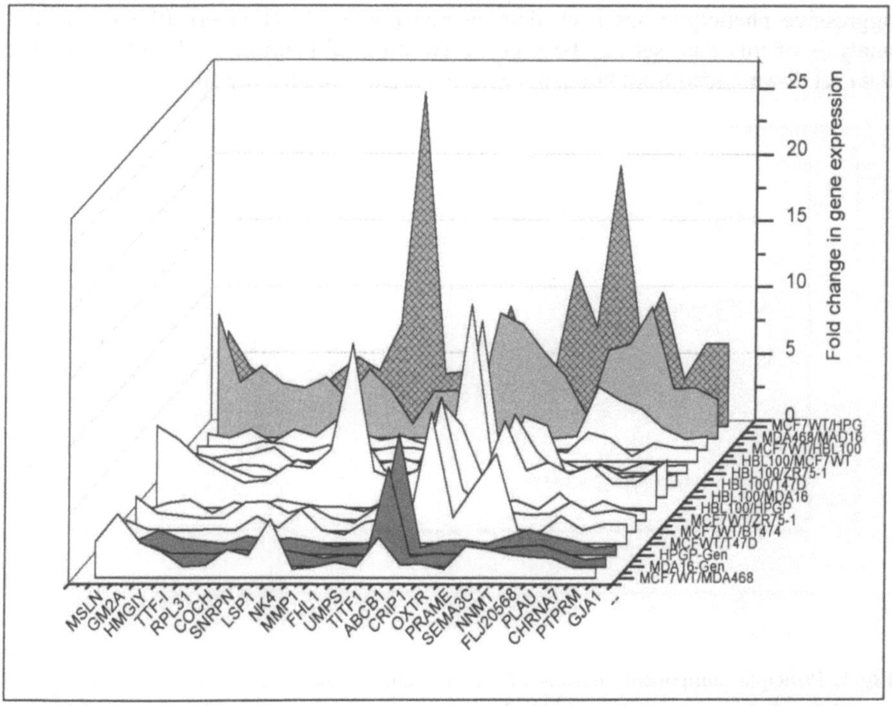

Fig 3. Gene expression profile of the drug resistant cluster from Turton *et al* (Turton et al. 2001). Shown are the genes that clustered with expression of the *ABCB1* gene in both cells with intrinsic and acquired resistance. On the y axis is shown the fold change of the gene expressions shown on the x axis of the cells on the right of the pair (z axis) compared with those on the left of the pair(z axis). Shown in grey hatch are the acquired doxorubicin resistant HPGP cells compared with their drug sensitive progenitors and in the solid light grey the intrinsically resistant MDA16 cells compared to their drug sensitive progenitors. In the darker grey are shown two hybridisations that are not of cDNA but of genomic DNA. These were performed to indicated gene amplifications and deletions and labelled using a random primer technique (Turton et al. 2001). The two cell lines shown are the HPGP and MDA16 cells both of which show an amplification of the *ABCB1* gene.

Using a principal components analysis the cell pairs could be sorted on the basis of their differential gene expression for similarity. With this analysis the cell pairs of both the acquired and intrinsically resistant cells grouped together when compared against a common reference (HBL100). When compared against their drug sensitive progenitors there was still similarity in terms of the MDR cluster, but the sets could be separated as a result of the gene expression differences between the parental drug sensitive cell lines. Similarly when the MDR cluster was not present, that is when the drug sensitive progenitor cells were compared against a common reference (HBL100) then these cells could also be separated (Fig 4). The MDR gene cluster also indicates other potential properties of the cells, for example the propensity towards a more metastatic phenotype indicated by the over-expression of *MMP1*. The propensity of MDR cells to have a more aggressive phenotype has been documented previously (Lehnert 1996). The full analysis of this data set can be seen in Turton *et al* (Turton et al. 2001) and the data set downloaded from http://www.le.ac.uk/cmht/twg1/array-fp.html.

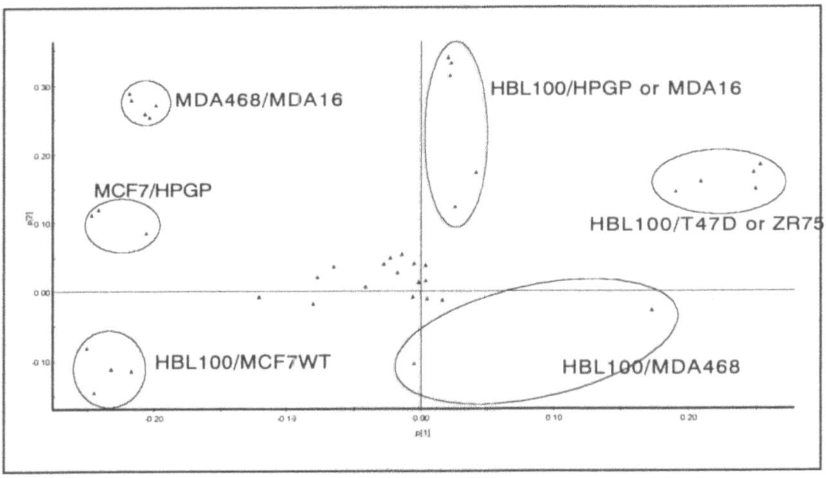

Fig 4. Principal component analysis of doxorubicin resistant cell lines. The entire sets of microarray expression data was subjected to a principal components analysis. The first two principal components are plotted against one another here and show how with this set of genes the acquired (HPGP) and intrinsically resistant (MDA16) cells are indistinguishable when compared against a common (HBL100) background. When compared against their drug sensitive progenitor cells then they can be differentiated because the differences between the cell lines overrides the similarity gained from the MDR gene cluster. Similarly the two progenitor drug sensitive cell lines can be separated when compared against the common HBL100 background, when the MDR profile is not present.

Conclusions

The data shown here serves to support the hypothesis that a tumor with an acquired resistance phenotype arises from the chemotherapy driven selection of cells from within the sensitive population. Intrinsically resistant cells are given a growth advantage by the selective killing of the drug sensitive cells and grow through to form the tumor with the acquired resistance phenotype. A tumor with acquired resistance is therefore derived from cells which have an intrinsic resistance phenotype derived from mutations, and are selected for in a Darwinian fashion through the administered chemotherapy.

References

Davies R, Budworth J, Riley J, Snowden R, Gescher A, Gant TW (1996) Regulation of P-glycoprotein 1 and 2 gene expression in two MCF-7/Adr cell line subclones. Brit J Cancer 73:307-315

Endicott JA, Ling V (1989) The biochemistry of P-glycoprotein mediated multidrug resistance. Ann Rev Biochem 58:137-171

Goldie JH, Coldman AJ, Gudauskas GA (1982) Rationale for the use of alternating non-cross-resistant chemotherapy. Cancer treatment reports 66:439-449

Harrison DJ (1995) Molecular mechanisms of drug resistance in tumours. J Path 175:7-12

Hayes JD, Wolf CR (1990) Molecular Mechanisms of Drug Resistance. Biochem J 272:281-295

Lehnert M (1996) Clinical multidrug resistance in cancer: A multifactorial problem. European Journal of Cancer Part A 32:912-920

Tsuruo T, Hamada H, Sato S, Heike Y (1989) Inhibition of multidrug resistant tumour growth in athymic mice by anti-P-glycoprotein monoclonal antibodies. Jpn J Cancer Res (Gann) 80:627-631

Turton NJ, Judah DJ, Riley J, Davies R, Lipson D, Styles JA, Smith AG, Gant TW (2001) Gene expression and amplification in breast carcinoma cells with intrinsic and acquired doxorubicin resistance. Oncogene 20:1300-1306

Vasudevan S, Tsuruo T, Rose DR (1998) Mode of binding of anti-P-glycoprotein antibody MRK-16 to its antigen - A Crystallographic and molecular modeling study. J Biol Chem 273:25413-25419

Weinstein RS, Grogan TM, Kuszak JR, Jakate SR, Kluskens LF, Coon JS (1991) Multidrug resistance gene product (P-glycoprotein) in normal tissue and tumors: Advances in pathology and laboratory medicine. Mosby year end book, Vol 6, pp 207-234

Screening trial of cancer chemopreventive agent by using combination of DNA chip technology and Cell Signaling Network DataBase (CSNDB)

Eiichiro Ichiishi and Toshikazu Yoshikawa

1st Dept of Medicine, Kyoto Prefectural University of Medicine, Kajii-cho, Kamigyo-ku, Kyoto 602-8566, Japan

Summary. Mechanism of the cancer prevention was complicated, and the elucidation was difficult if it was a conventional experiment system. We are building the system which put the latest technology to reach a screening trial of cancer preventive agents from an approach to find candidate cell signaling pathway of cancer prevention by using combination of DNA chip technology and Cell Signaling Network Database (CSNDB) maintained by the National Institute of Health Sciences. As for these strategies, we think that it is an interesting trial when we examine the effectiveness from an interesting pathway search in investigation for an unknown cancer preventive agent. We are in the way, but demonstrate a part of our study and try to think about the future prospects.

Keywords. Chemopreventive agent, DNA chip, drug development, cell signaling network

Introduction

Many inhibitors of carcinogenesis promoters have been identified by the EB virus early antigen expression inhibition test on Raji cells [1 - 6], and their effects have been confirmed mostly by the mouse skin two-step carcinogenesis inhibition test *in vivo* [7 - 11]. However, gene analysis associated with the EB virus early antigen expression inhibition test on Raji cells has not fully investigated.

The DNA chip microarray technique that has rapidly developed in recent years allows systematic screening of gene expression [12,13]. Previous experiment of aging and apoptosis using microarrays of cultured cells provided very significant data [14]. DNA chips may become the core technique facilitating genome research in the EB virus early antigen expression inhibition test, but difficulties in reciprocal comparison and interpretation of data are obstacles to the study. In the present study, we investigated the effect of retinoic acid on gene expression on Raji cells induced by 12-*O*-tetra-decanoylphorbol-13-acetate (TPA). In addition, as one trial, we compared a large amount of data in DNA chips with cell signaling network database (CSNDB) developed at the National Institute of Health Sciences and tried to analyze partly the role of genes in intracellular signal transmission [15,16]. In this strategy, we seek the screening trial of cancer chemopreventive agent by using combination of DNA chip technology and CSNDB.

Materials and Methods

Chemicals

The cell culture reagents, n-butyric acid and other reagents for EBV-EA assay were purchased from Nakarai Tesque Inc.(Tokyo) High grade solvents (Sigma, St.Louis) were used for the superoxide scavenging assay and 7,12-dimethyl-benz(a)anthracene (DMBA) and 12-*O*-tetra-decanoylphorbol-13-acetate (TPA) for the *in vivo* assay.

Mouse skin test

The experiment was performed by the standard procedure [17,18]. Groups of 15 ICR female mice [The SLC Inc. (Shizuoka, Japan)] were used. The skin of the back of all of the mice was shaved with an electric clipper one day prior to the initiator treatment. Initiation was accomplished by a single topical treatment with 100 microgram of DMBA on shaved backs. Application of 1 microgram of TPA was started one week after initiation, twice a week. 8.5×10-5M of retinoic acid was applied topically one hour before each TPA application (pre-treatment) and the promoting phase carried out for 20 weeks. The dose level of retinoic acid against TPA was set to 50 times by volume weight. Skin papilloma formation was recorded weekly in each experimental group.

Cell preparation and DNA chip technology

EA positive and negative Raji cells were prepared in the previous method [2]. 32 nM administration of retinoic acid on Raji cells dramatically inhibited EA induction [2]. Affymetrix oligonucleotide arrays were used for mRNA expression profiling [12 - 14]. Polyadenylated RNA was isolated from 5×10^7lysed Raji cells of EA expression positive and negative. Once cDNA was synthesized and biotinylated cRNA was generated as previously described [19]. Fragmented biotin-cRNAs (50 microgram) were hybridized to the perfect match (PM) and mismatch (MM) oligonucleotides on the Hu6800 GeneChip (Affymetrix). Digitized image data were processed using the GeneChip software (version 3.0) from Affymetrix as described previously [14]. Gene induction or repression was considered significant if the change in average difference intensity was above 2-fold or greater changes were scored as significant changes [20].

Cell Signaling Networks Data Base

Cell Signaling Networks DataBase (CSNDB) is a data- and knowledge- base for signaling pathways of human cells. Signaling pathways are compiled as binary relationships of biomolecules and represented by graphs drawn automatically in the form of Accession (AC) numbers. The number of the information of binary relationships of biomolecules is over 1200 and over 1400 references have been quoted. CSNDB has a linkage to The Transcription Factor Database (TRANSFAC) [15,16].

The CSNDB web site is as follows; http://geo.nihs.go.jp/csndb/

Results

Mouse skin test

The *in vivo* inhibitory activity was tested by two-stage mouse skin carcinogenesis using DMBA as the initiator and TPA as the promoter. We demonstrated the inhibition of papilloma formation in the case when the 8.5×10-5M of retinoic acid was applied one hour after the TPA treatment. In mice treated with retinoic acid, the formation of papillomas was observed to be delayed for 1 week and resulted in a 60% reduction in the number of papillomas per mouse.

DNA chip expression analysis and comparing with CSNDB

In an attempt to identify genes whose expression changed during 32 nM retinoic acid showed inhibitory effect of EA, we utilized DNA chip technology to obtain an overall profile of gene expression.

Since differential expression analysis between EA positive cells and EA inhibited cells are demonstrated, EA inhibitory effect related genes by retinoic acid are expected to be identified. In order to identify both EA positive and EA negative genes, poly mRNA was isolated from cells treated with or without retinoic acid for 8 h and subjected to oligonucleotide chip hybridization. As compared with two chips, containing 6591 genes were screened. The hybridization results from two chips were compiled and sorted on the basis of differential call and fold change compared to control cells. Genes that displayed approximately two-fold or greater changes were scored as significant changes [20]. 322 mRNA species out of the 6591 genes were identified by these criteria (data not shown). Down-regulated genes are 109 and up-regulated genes are 213(data not shown).

In these 322 genes, based on the GenBank AC numbers, we collated the data with CSNDB and 19 molecules were consistent of CSNDB program. The names of these molecules are listed in Table 1. Up-regulated genes are 11 and down are 8 species.

Discussion

In this study, we used retinoic acid which is a typical EA inhibitor [2]. Because retinoic acid is currently expected for application to medical care practice such as differentiation induction therapy for leukemia [21], we selected it as a candidate substance expected for clinical success. Retinoic acid has important roles in regulating genes involved in cell morphogenesis, differentiation, and proliferation [22]. Variety function of retinoic acid may explain the dramatically effect of tumour prevention *in vivo*. Retinoid-responsible genes are considered to be complexity and further more investigations remained [22].

In the present study, as a large number of genes were systemically demonstrated by DNA chip analysis, we tried to pick up genes by using CSNDB program according to function or signal transmission. CSNDB program contains various information about signaling pathways and important molecules of physiological dynamics are almost included in the form of GeneBank AC Numbers, however, CSNDB program is now under construction and not a mature system [15,16]. In the present study, we have found that the result of all 19 genes in Fig.1 on Raji cells by administrating retinoic acid are new findings. These 19 genes partially may account for the functions of retinoic acid but other famous signals including RAR or RXR pathways [22] were not extracted. This reason may be explained by following two speculations that very low concentration of 32 nM retinoic acid affects Raji cells insufficiently and developing system of CSNDB works incompletely not to extract useful data.

Among genes up-regulated, ubiquitin gene has been reported to be induced on human keratinocytes by retinoic acid [23]. Many short-lived regulatory proteins are degraded by the ubiquitin/proteasome pathway [23]. Several lines of evidence have indicated that the proteasome has the major proteolytic activity responsible for generation of antigenic peptides [24]. It may be possible that ubiquitin/proteasome pathway acts as proteolysis of early antigen protein.

We also observed that differentiation-related genes were up-regulated. JunD and FYN oncogene have been proposed to be the important factor of differentiation as to neuroblastoma and oligodendrocytes, respectively [25,26]. It is very interesting that retinoic acid, an inhibitor of carcinogenesis promotor, induces JunD and FYN gene, differentiation-related genes. Clathlin is also related to postsynaptic differentiation [27]. Receptor-related genes including TNF receptor-associated factor 4 (TRAF4), interleukin-1 receptor-associated kinase1 and vasopressin V3 receptor genes were also induced. The significance of up-regulation of receptor-related genes are unknown in the present study. It may be possible that autocrine

Table 1

Probe Set	CSNDB	Differential Call	Fold Change	Accession Numbers	Gene
U49869_rna1	ubiquitin	I	2.5	U49869_rna1	ubiquitin
X56681_s_at	JunD	I	1.5	X56681_s	jun D proto-oncogene
M20471_at	clathrin	I	1.5	M20471	clathrin, light polypeptide (Lca)
X02317_at	superoxide dismutase	I	1.5	X02317	superoxide dismutase 1, soluble (amyotrophic lateral sclerosis 1 (adult))
X80200_at	TRAF4	I	2	X80200	TNF receptor-associated factor 4
L11285_at	MEK2	I	1.2	L11285	protein kinase, mitogen-activated, kinase 2, p45 (MAP kinase kinase 2)
L76191_at	IRAK	I	1.3	L76191	interleukin-1 receptor-associated kinase 1
M90299_at	MODY2	I	~2.8	M90299	glucokinase (hexokinase 4, maturity onset diabetes of the young 2)
M74491_at	ARF	I	~2.5	M74491	ADP-ribosylation factor 3
L37112_at	vasopressin receptor	I	~2.5	L37112	vasopressin V3 receptor
M14676_at	Fyn	I	~2.0	M14676	FYN oncogene related to SRC, FGR, YES
M33680_at	CD81	D	-1.2	M33680	CD81 antigen (target of antiproliferative antibody 1)
X95404_at	cofilin	D	-1.5	X95404	cofilin 1 (non-muscle)
M30448_s_at	casein kinase 2	D	-1.3	M30448_s	casein kinase 2, beta polypeptide
L20688_at	rhoGDI	D	-1.2	L20688	Rho GDP dissociation inhibitor (GDI) beta
X53416_f_at	actin	D	-4.1	X53416_f	actin, F
M19311_s_at	calmodulin	D	-2	M19311_s	calmodulin 1 (phosphorylase kinase, delta)
D87116_at	MAP-kinase-kinase	D	-1.4	D87116	protein kinase, mitogen-activated, kinase 3 (MAP kinase kinase 3)
J03801_f_at	lysozyme	D	-2.7	J03801_f	lysozyme (renal amyloidosis)

I:Increase
D:Decrease

or paracrine regulations through receptor induction are functioned by retinoic acid. P45 MAPK gene has been reported to be induced by IL5 on the hematopoietic cells. This isoform of MAPK also has considered to relate to differentiation [28].

As to down-regulated genes, CD81 (TAPA-1) is a widely expressed cell-surface protein involved in an astonishing variety of biologic responses related to cell adhesion, motility, and metastasis [29]. Physiological significance of down-regulation of CD81 needs further more investigation. Casein kinase II is a serine/threonine kinase over-expressed in many human tumors, transformed cell lines, and rapidly proliferating tissues [30]. In this study, inhibition of Casein kinase II gene on Raji cell may be a new and interesting function of retinoic acid.

One of our trial studies is to investigate how to extract candidate pathways efficiently from enormous gene expression data. This trial may be invaluable for DNA chip/microarray users because of the new methodology of combined array data and database information. We put the data of 19 species in Fig.1 into CSNDB program and extracted three pathways.

From "rhoGDI" to "actin" (Fig 1)

This pathway contains the DNA chip data of co-suppression of "cofilin" and "actin"in contrast of Table.1. Increased expression of cofilin was reported to the candidate marker of drug-resistant human adenocarcinoma of the pancreas [31]. It can be possibly said that cofilin-mediated pathways may be promotion- or progression-related signals of the poor-prognosis cancer and inhibition of cofilin signal by retinoic acid is of interest.

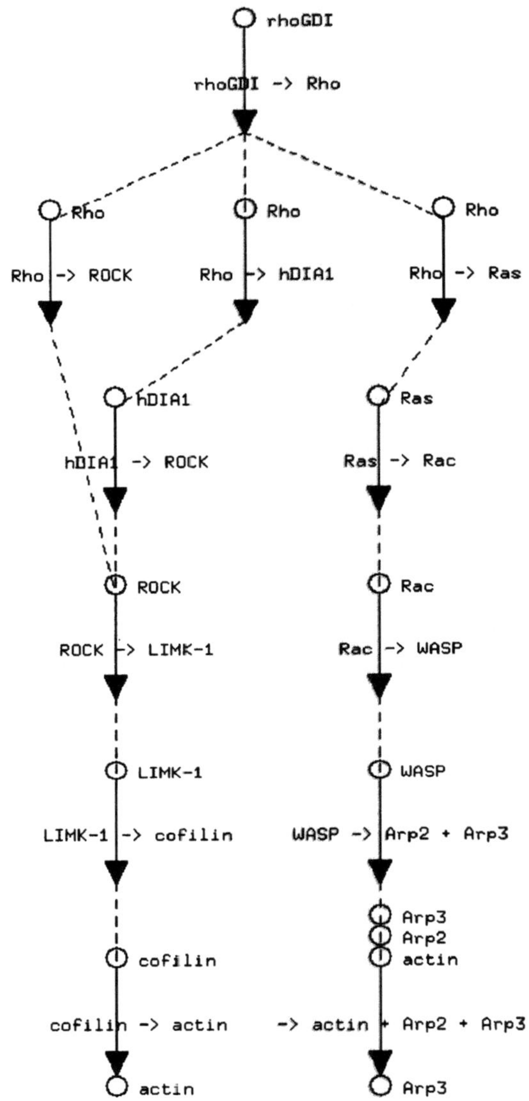
rhoGDI
rhoGDI -> Rho
Rho
Rho
Rho
Rho -> ROCK
Rho -> hDIA1
Rho -> Ras
hDIA1
Ras
hDIA1 -> ROCK
Ras -> Rac
ROCK
Rac
ROCK -> LIMK-1
Rac -> WASP
LIMK-1
WASP
LIMK-1 -> cofilin
WASP -> Arp2 + Arp3
Arp3
Arp2
cofilin
actin
cofilin -> actin
-> actin + Arp2 + Arp3
actin
Arp3

Fig. 1

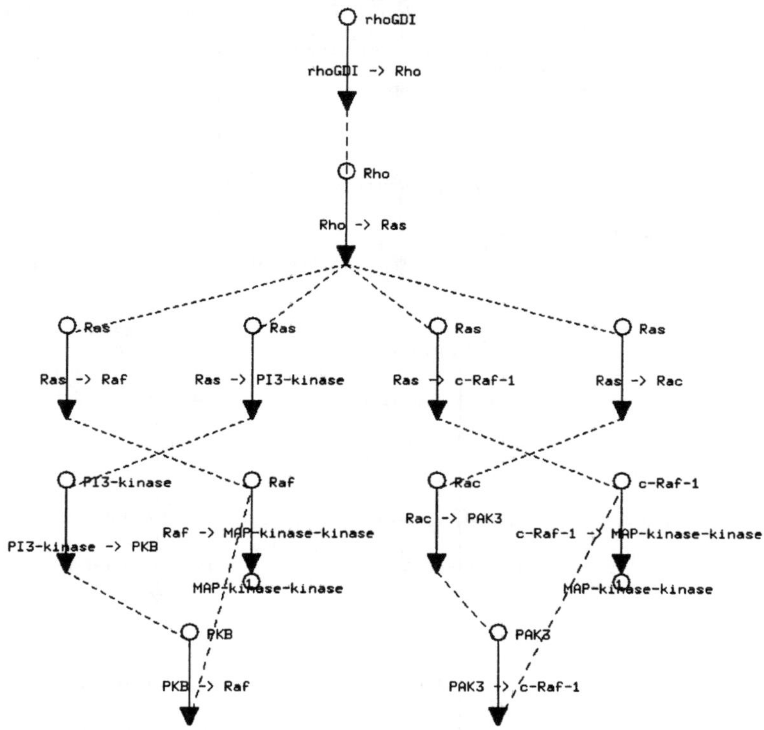

Fig. 2

From "rhoGDI" to "MAP-kinase-kinase" (Fig 2)

This pathway contains the DNA chip data of co-suppression of "rhoGDI" and "MAP-kinase-kinase of Table 1. Rho family are involved in the regulation of a variety of cellular processes, such as the organization of the microfilamental network, cell-cell contact and malignant transformation [32]. Modulation of this signaling pathway by retinoic acid may have significant information related to carcinogenesis.

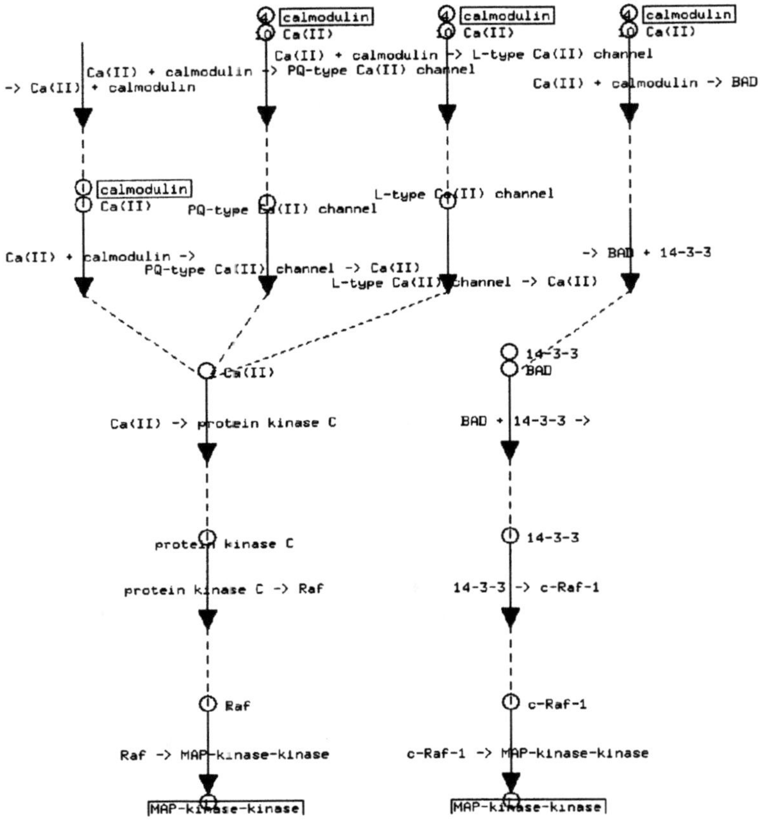

Fig. 3

From "calmodulin" to "MAP-kinase-kinase" (Fig 3)

This pathway contains the DNA chip data of co-suppression of "calmodulin" and "MAP-kinase-kinase" of Table 1. Calmodulin mediated pathway is interesting for cancer prevention and antagonist of calmodulin has been to paid attention[33]. Retinoic acid may work as an antagonist of calmodulin in the level of gene expression.

From this study, some up-regulation genes on Raji cells by retinoic acid may play roles in induction of differentiation in various style and down-regulation genes may related to tumour specific genes which may be one of the therapeutic target from a view to pharmacogenomics [34].

In this strategy, we are tring to find a useful cancer chemopreventive agent by candidate pathway of cancer prevention.

Rapid and simultaneous analysis of multiple parameters and their interaction using computer databases as done in this study, may elucidate complex biosystems and provide clues that will greatly facilitate not only the medical and pharmaceutical field but also exploration of unknown areas in various fields.

Acknowlegements

We gratefully acknowledge the expert assistance of Dr.Tsuguchika Kaminuma and Dr.Takako Takai in DNA chip data analyses by using Cell Signaling Network DataBase.

References

1. zur Hausen H, Bornkamm GW, Schmidt R, Hecker E (1979) Tumor initiators and promoters in the induction of Epstein-Barr virus. Proc Natl Acad Sci USA 76:782-785
2. Yamamoto N, Bister K, zur Hausen H (1979) Retinoic acid inhibition of Epstein-Barr virus induction. Nature 278:553-554
3. zur Hausen H, O'Neill F, Freese U, Hecker E (1978) Persisting oncogenic herpesvirus induced by the tumour promotor TPA. Nature 272:373-375
4. Konoshima T, Takasaki M, Ichiishi E, Murakami T, Tokuda H, Nishino H, Duc NM (1999) Cancer chemopreventive activity of majonoside-R2 from Vietnamese ginseng, *Panax vietnamensis*. Cancer Lett 147:11-16
5. Nishino H, Murakoshi M, Masuda M, Tokuda H, Satomi Y, Onozuka M, Yamaguchi S, Bu P, Tsuruta A, Nosaka K, Baba M, Takasuka N (1999) Suppression of lung and liver carcinogenesis in mice by oral administration of myo-inositol. Anticancer Res 19:3663-3664
6. Konoshima T, Takasaki M, Tokuda H (1999) Anti-carcinogenic activity of the roots of Panax notoginseng. II. Biol Pharm Bull 22:1150-1152
7. Colombo D, Compostella F, Ronchetti F, Scala A, Toma L, Kuchide M, Tokuda H, Nishino H (2000) Anti-tumor-promoting effects of glycoglycerolipid analogues on two-stage mouse skin carcinogenesis. Cancer Lett 20:201-205
8. Tanaka R, Ohtsu H, Iwamoto M, Minami T, Tokuda H, Nishino H, Matsunaga S, Yoshitake A (2000) Cancer chemopreventive agents, labdane diterpenoids from the stem bark of *Thuja standishii (Gord.) Carr*. Cancer Lett 20:165-170
9. Ito H, Koreishi M, Tokuda H, Nishino H, Yoshida T (2000) Cypellocarpins A-C, phenol glycosides esterified with oleuropeic acid, from *Eucalyptus cypellocarpa*. J Nat Prod 63:1253-1257
10. Ishida J, Kozuka M, Wang H, Konoshima T, Tokuda H, Okuda M, Yang Mou X, Nishino H, Sakurai N, Lee KH, Nagai M (2000) Antitumor-promoting effects of cyclic diarylheptanoids on Epstein-Barr virus activation and two-stage mouse skin carcinogenesis. Cancer Lett 31:135-140

11. Iwase Y, Takemura Y, Ju-ichi M, Ito C, Furukawa H, Kawaii S, Yano M, Mou XY, Takayasu J, Tokuda H, Nishino H (2000) Inhibitory effect of flavonoids from citrus plants on Epstein-Barr virus activation and two-stage carcinogenesis of skin tumors. Cancer Lett 154:101-105
12. Lockhart DJ, Dong H, Byrne MC, Follettie MT, Gallo MV, Chee MS, Mittmann M Wang C, Kobayashi M, Horton H, Brown EL (1996) Expression monitoring by hybridization to high-density oligonucleotide arrays. Nat Biotechnol 14:1675-1680
13. Wodicka L, Dong H, Mittmann M, Ho MH, Lockhart DJ (1997) Genome-wide expression monitoring in *Saccharomyces cerevisiae*. Nat Biotechnol 15:1359-1367
14. Wang Y, Rea T, Bian J, Sun Y (1999) Identification of the genes responsive to etoposide-induced apoptosis: application of DNA chip technology. FEBS Lett 445:269-273
15. Takai-Igarashi T, Nadaoka Y, Kaminuma T (1998) A database for cell signaling networks. J Comput Biol 5:747-754
16. Igarashi T, Kaminuma T (1997) Development of a cell signaling networks database. Pac Symp Biocomput :187-197
17. Belman S, Troll W (1972) The inhibition of croton oil-promoted mouse skin tumorigenesis by steroid hormones. Cancer Res 32:450-454
18. Verma AK, Slaga TJ, Wertz PW, Mueller GC, Boutwell RK (1980) Inhibition of skin tumor promotion by retinoic acid and its metabolite 5,6-epoxyretinoic acid. Cancer Res 40:2367-2371
19. Der SD, Zhou A, Williams BR, Silverman RH (1998) Identification of genes differentially regulated by interferon alpha, beta, or gamma using oligonucleotode arrays. Proc Natl Acad Sci USA 95:15623-15628
20. de Saizieu A, Certa U, Warrington J, Gray C, Keck W, Mous J (1998) Bacterial transcript imaging by hybridization of total RNA to oligonucleotide arrays. Nat Biotechnol 16:45-48
21. Huang ME, Ye YC, Chen SR, Chai JR, Lu JX, Zhoa L, Gu LJ, Wang ZY (1988) Use of all-trans retinoic acid in the treatment of acute promyelocytic leukemia. Blood 72:567-572
22. Hurst RE, Waliszewski P, Waliszewska M, Bonner RB, Benbrook DM, Dar A, Hemstreet GP (1999) Complexity,retinoid responsive gene networks, and bladder carcinogenesis. Adv Exp Med Biol 462:449-467
23. Boudjelal M, Wang Z, Voorhees JJ, Fisher GJ (2000) Ubiquitin/proteasome pathway regulates levels of retinoic acid receptor gamma and retinoid X receptor alpha in human keratinocytes. Cancer Res 15:2247-2252
24. York IA, Goldberg AL, Mo XY, Rock KL (1999) Proteolysis and class I major histocompatibility complex antigen presentation. Immunol Rev 172:49-66
25. de Groot RP, Kruijer W (1991) Up-regulation of Jun/AP-1 during differentiation of N1E-115 neuroblastoma cells. Cell Growth Differ 2:631-636
26. Osterhout DJ, Wolven A, Wolf RM, Resh MD, Chao MV (1999) Morphological differentiation of oligodendrocytes requires activation of Fyn tyrosine kinase. J Cell Biol 145:1209-1218
27. Bloch RJ, Pumplin DW (1988) Molecular events in synaptogenesis: nerve-muscle adhesion and postsynaptic differentiation. Am J Physiol 254:C345-364
28. Bates ME, Bertics PJ, Busse WW (1996) IL-5 activates a 45-kilodalton mitogen-activated protein (MAP) kinase and Jak-2 tyrosine kinase in human eosinophils. J Immunol 156:711-718

29. Levy S, Todd SC, Maecker HT (1998) CD81 (TAPA-1): a molecule involved in signal transduction and cell adhesion in the immune system. Annu Rev Immunol 16:89-109
30. Song DH, Sussman DJ, Seldin DC (2000) Endogenous protein kinase CK2 participates in Wnt signaling in mammary epithelial cells. J Biol Chem 275:23790-23797
31. Sinha P, Hutter G, Kottgen E, Dietel M, Schadendorf D, Lage H (1999) Increased expression of epidermal fatty acid binding protein, cofilin, and 14-3-3-sigma (stratifin) detected by two-dimensional gel electrophoresis, mass spectrometry and microsequencing of drug-resistant human adenocarcinoma of the pancreas. Electrophoresis 20:2952-2960
32. Fritz G, Just I, Kaina B (1999) Rho GTPases are over-expressed in human tumors. Int J Cancer 81:682-687
33. Seiler N, Atanassov CL, Raul F (1998) Polyamine metabolism as target för cancer chemoprevention. Int J Oncol 13:993-1006
34. Tanaka T, Nishimura Y, Tsunoda H, Naka M (2000) Pharmacogenomics and therapeutic target validation in cerebral vasospasm. J Cardiovasc Pharmacol 36:Suppl 2, S1-4

Application of methacarn fixation for genetic analysis in microdissected paraffin-embedded tissue specimens

Makoto Shibutani, Chikako Uneyama, Naoya Masutomi, Hironori Takagi, and Masao Hirose
Division of Pathology, National Institute of Health Sciences, 1-18-1, Kamiyoga, Setagaya-ku, Tokyo 158-8501, Japan

Summary. Recent advances in microdissection have succeeded the molecular biological approaches on specific cells of interest within tissue specimens by overcoming the obstacle of tissue complexity. Recently, we found that methacarn is suitable for analysis of RNA, protein, and genomic DNA in small tissue samples using paraffin-embedded tissue (PET) sections in conjunction with microdissection technique. By application of sensitive quantitation methods, such as those utilizing fluorescent dyes specific for RNA or protein, molecules of small quantity can be normalized between samples, and thus quantitative expression analysis for RNA or protein can be applied in microdissected small tissue specimens. In addition, methacarn-fixation extends its availability for genomic DNA analysis in terms of target fragment size and number of microdissected cells required. Paraffin embedding permits ease of handling tissues that extend the availability of methacarn fixation for genetic analysis in large-scale experiments. In addition, considering its advantages for immunohistochemistry, tissue embedding after methacarn-fixation should be recommended as a valuable approach for routine application possibly in combination with targeted genetic analysis of immunophenotypically defined cell populations. In combination with techniques such as expression library construction, microarray and subtractive hybridization or differential display, microdissection will permit the establishment of "genetic fingerprints" of specific cellular areas.

Key words. Microdissection, Methacarn fixation, Gene expression, Mutation analysis, Paraffin-embedded tissue

Introduction

Analysis of pathological tissue specimens in organ at the level of RNA, protein and DNA has contributed to the investigation of molecular events as well as the advance of diagnostics. However, the cellular heterogeneity composing organs/tissues with an admixture of reactive cell populations may influence the outcome of analyzed results. Recent advances in microdissection have succeeded the molecular biological approaches on specific cells of interest within tissue specimens (Emmert-Buck et al. 1996; Schütze and Lahr 1998). Quantitative gene expression analysis in the pathologically altered cell populations can provide valuable information regarding the mechanism underlying biological phenomena, such as inflammatory cell response, cell growth, differentiation, and apoptotic cell death. Furthermore, mutation analysis of single cells becomes now essential for investigation of carcinogenetic mechanisms.

For histological assessment, tissue fixation and subsequent paraffin embedding are routinely employed because of the ease of handling tissues and subsequent staining, as well as the good preservation of tissue architecture. Until recently, formaldehyde-based fixatives, such as buffered-formalin, have been used for this purpose. However, with such cross-linking agents, there is limited performance in terms of the efficiency of extraction and quality of extracted RNA (Cooms et al. 1999; Foss et al. 1994), protein (Ikeda et al. 1998; Shibutani et al. 2000), and genomic DNA (Berger et al. 1999; Blomeke et al. 1997; Diaz-Cano and Brady 1997; Foss et al. 1994; Frank et al. 1996; Howe et al. 1997; Merkelbach et al. 1997; Ortiz Pallardo et al. 2000; Poncin et al. 1999; Scholte et al. 1997; Whittington et al. 1999), with consequent difficulty in the analysis of microdissected, histologically defined tissue areas. Therefore, unfixed tissue preparation now becomes the goldstandard for analysis of microdissected cells. However, preparation of cryosections from unfixed frozen tissue for the purpose of microdissection may not be suitable for routine samples because of the inconvenience in terms of tissue storage and the skill required for cryosection preparation and subsequent microdissection. As compared to unfixed frozen tissue, paraffin-embedded tissue (PET) retains benefits in terms of the performance in handling tissues of many samples and accessibility to histologically defined cells such as specific neuronal cell populations.

For molecular analysis of microdissected cells, extraction efficiency and quality of molecules are critical. Recently, we found that methacarn meets these critical criteria for analysis of RNAs, proteins and genomic DNAs in defined areas of PET sections (Shibutani et al. 2000; Uneyama et al. 2000), Methacarn is a non-cross-linking protein-precipitating fixative, and has been developed as a methanol-based Carnoy's solution to reduce tissue shrinkage by ethanol (Puchtler et al. 1970). As the main focus of this article, we present procedures of methacarn fixation and following quantitative gene expression analysis and genomic DNA analysis applicable for microdissected small cell areas in PET sections.

Fixation and Paraffin Embedding

Methacarn solution consisting of 60% (vol/vol) absolute methanol, 30%

chloroform, and 10% glacial acetic acid should freshly be prepared and stored at 4°C before fixation. Organs/tissues of interest can be trimmed 3~4 mm in thickness to facilitate fixation. Tissues are fixed with methacarn for 2 hr at 4°C. For embedding, tissue samples are dehydrated 3 times for 1 hr in fresh 100% ethanol at 4°C, immersed in xylene once for 1 hr and then 3 times for 30 min at room temperature, and immersed in hot paraffin (60°C) 3 times for 1 hr.

Microdissection

There are two major techniques for microdissection utilizing the precision of lasers. One technique is laser microbeam microdissection (LMM) with laser pressure catapulting to recover whole targeted cell area of any size, and this system is based on a pulsed UV laser with a small beam focus to cut out areas or cells of interest by photoablation of adjacent tissue. Another technique is laser capture microdissection (LCM), which utilizes a low energy infrared laser pulse to capture the targeted cells by focal melting of the thermoplastic membrane through laser activation. Advantages and disadvantages of these systems are described elsewhere (Fend and Raffeld, 2000).

In our laboratory, we use LMM with laser pressure catapulting by PALM® Robot-Microbeam equipment (Carl Zeiss Co., Ltd., Tokyo, Japan) because whole targeted cell area of any size can be recovered with this system. Briefly, tissue sections from PET, 4-30 mm in thickness, are prepared and mounted onto 1.35 mm thin polyethylene film, deparaffinized with xylene 3 times for 10 min and placed in 100% ethanol for washing 3 times for 10 min. If it is necessary, tissue section can briefly be stained with hematoxylin or cresyl violet. Then the membrane with the attached specimen is mounted in reverse (membrane side up) onto a new cover slip. The targeted tissue area is then subjected to microbeam microdissection and removed either by a needle tip or laser pressure catapulting, depending on the size of tissue fragments microdissected.

RNA analysis

Microdissected tissue fragments are lysed with RNA STAT-60 (Tel-Test "B", Inc., Friendswood, TX) and total cytoplasmic RNA is isolated with isopropanol precipitation in the presence of glycogen at 2 mg/ml as a carrier. Isolated RNA is then labeled with a RiboGreen™ RNA Quantitation kit (Molecular Probes, Eugene, Oregon) and concentrations are estimated with a fluorescence spectrophotometer.

In our laboratory, we found that methacarn-fixed rat liver PET allows the extraction of total RNA yielded 52 ± 15 ng/mm^2, sufficient for quantitative RT-PCR of many genes, from a deparaffinized 10 mm thick section by a simple, single step extraction method (Shibutani et al. 2000). When deparaffinized small tissue blocks were examined for extraction, methacarn-fixed rat liver PET showed 46% of RNA yield of the unfixed frozen tissue, and it was superior to any of the cross-linking fixatives examined (Shibutani et al. 2000).

Reverse transcription-polymerase chain reaction (RT-PCR)

Single strand DNA is generated from 15 ng ~1000 ng of total RNA with random hexamers and the SUPERSCRIPT™ Preamplification System (Life Technologies, Inc., Rockville, MD). Hot start PCR is performed with PLATINUM *Taq* DNA polymerase (Life Technologies, Inc.) in a total 20-ml volume. The cycle parameters for the PCR of target fragment sized smaller than 1 kilobase (kb) are typically 94°C for 1 min, 55°C for 1 min, and 72°C for 1 min.

For analysis of RNA expression, purity and quality of extracted RNA must be ensured. We have already demonstrated that the integrity of total RNA extracted from methacarn-fixed paraffin-embedded rat PC12 cells was well preserved and similar to that from unfixed frozen cells, judging from the resolution of 18S and 28S ribosomal RNAs (Shibutani et al. 2000). Furthermore, the total RNA extracted from methacarn-fixed PET showed low concentration of contaminating genomic DNA as compared to those from acetone-fixed PET or unfixed frozen tissue. Successful PCR-amplification of long mRNA sequence (~1.9 kb) and any mRNA species, even those expressing low-copy numbers proved the quality of extracted RNA from methacarn-fixed PET sections (Shibutani et al. 2000).

Copy numbers of the target gene transcript in microdissected cells could be estimated by competitive RT-PCR and following plate hybridization (Shibutani and Uneyama, in press). Isolated total RNA from microdissected cells is reverse-transcribed to cDNA. PCR is performed with 5'-biotinylated upstream primer and unlabeled downstream primer. The resulting amplicons are detected with streptavidin-horse radish peroxidase (HRP) conjugate and 3,3',5,5'-tetramethylbenzidine as a substrate.

For competitive RT-PCR, preliminary validation study should be performed with the practical RNA samples on the concentration of competitor fragment, and cycle parameters including its numbers. Similarly, concentration of amplicons should be set excess over the background binding level due to the cross hybridization of the target amplicons (or competitor amplicons) to the competitor capture wells (or target wells). With 150~250 ng of total RNA isolated from microdissected cells, competitive RT-PCR can be performed. If equipment is available, real-time RT-PCR could also be applied.

Protein analysis

For protein extraction, tissue fragment is solubilized and sonicated for 10 min in

2x sodium dodesyl sulphate (SDS) gel loading buffer. If the mobility shift of the molecule of interest appears in the SDS-polyacrylamide gel electrophoresis (PAGE) by Western blotting, protein precipitation with 10% trichloroacetic acid (TCA) in saline should be performed before solubilization in 2x SDS buffer. Protein concentrations are estimated with a NanoOrange™ Protein Quantitation kit (Molecular Probes).

In terms of protein extraction, tissue section from methacarn-fixed PET can easily be solubilized in 2x SDS gel loading buffer. We found that extraction of protein yielded 4.9 ± 2.1 mg/mm^2 from a 10-mm thick rat liver section, allowing a quantitative expression analysis of protein by Western blotting. On the other hand, methacarn-fixed rat liver PET showed 77% of the protein yield of the unfixed frozen tissue (Shibutani et al. 2000). Although the so called "diffusion artifact" by such alcohol-based fixative may be responsible for the slight loss, signal intensities of membrane-bound proteins, such as epidermal growth factor receptor (EGFR) and cytochrome P450 (CYP) 2E1 did not differ between methacarn-fixed PETs and unfixed frozen samples, suggesting the loss in the protein yield could be negligible in analysis by Western blotting (Shibutani et al. 2000).

Protein samples are subjected to SDS-PAGE and then transferred to a polyvinylidine difluoride membrane (Millipore, Bedford, MA). After blocking with 0.2% casein (Merck, Whitehouse Station, NJ), blots are incubated with the primary antibody against the molecules of interest. Bound antibodies are detected with the secondary antibodies conjugated with HRP (Amersham Pharmacia Biotech AB, Uppsala, Sweden) and analyzed either with the Supersignal Chemiluminescent Substrate (Pierce Technology Corp. Inc., New York, NY) or the BIO-RAD HRP color reagent (Hercules, CA).

For protein analysis, non-crosslinking fixatives, such as acetone, ethanol or modified Carnoy's solution have been proven to give clear protein bands for cytokeratin molecules (Conti et al. 1988). We have also demonstrated that the polypeptide band of all examined proteins, in which two glycoproteins (cathepsin D and EGFR) were included, migrated to the positions of correct molecular size (17~170 kDa) in our methacarn-fixed PET. However, it should be noted that some unknown modification would appear on certain populations of polypeptides affecting the mobility in SDS-polyacrylamide gel without TCA precipitation of the tissue samples, such as the case of CYP2E1 signal. Therefore, it is recommended to perform TCA precipitation of microdissected cells before solubilization with 2x SDS buffer.

DNA analysis

For DNA extraction, microdissected cells are trapped in PCR tube caps and lysed with 4 ml of DEXPAT (Takara Shuzo Co. Ltd., Kyoto, Japan) at 95°C for 10 min, and the entire lysate is used as a template for PCR by adding to the master mix of total 50 ml directly. If it is necessary, DNA concentration in the final preparation can be measured using Hoechst 33258 (Molecular Probes).

In terms of genomic DNA analysis, methacarn-fixed PET showed approximately 8 times higher DNA recovery in unit area of 10-mm thick rat liver section as compared to that of buffered formalin-fixed PET (Uneyama et al. 2000).

Hot start PCR of the genomic sequence of the gene of interest is performed with PLATINUM *Taq* DNA polymerase in a 50-ml total reaction volume. If nested PCR is intended, 1 ml of first PCR product is used as a template in a 20 ml total volume. In the preliminary study, we have shown that the extensive portions of genomic DNA sequence, up to 4 kb could be amplified by single-step PCR with extracted DNA from the rat liver PET section as a template (Uneyama et al. 2000). Recently, we investigated the performance to amplify genomic DNA by nested PCR, and found that a DNA fragment of 522 base pairs could be amplified in 20% of microdissected single Purkinje cells from 10 mm rat cerebellar PET section, and the minimum number of cells required for practical PCR analysis as estimated using rat hippocampal neurons was in the order of 10-20 (manuscript in preparation).

Acknowledgments

This work was supported by a Grant-in-Aid from the Ministry of Health and Welfare of Japan (Grant H11-Seikatsu-20).

References

Berger DH, Chang H, Wood M, Huang L, Heath CW, Lehman T, Ruggeri B (1999) Mutational activation of K-ras in nonneoplastic exocrine pancreatic lesions in relation to cigarette smoking status. Cancer 85:326-332

Blomeke B, Bennett WP, Harris CC, Shields PG (1997) Serum, plasma and paraffin-embedded tissues as sources of DNA for studying cancer susceptibility genes. Carcinogenesis 18:1271-1275

Conti CJ, Larcher F, Chesner J, Aldaz CM (1988) Polyacrylamide gel electrophoresis and immunoblotting of proteins extracted from paraffin-embedded tissue sections. J Histochem Cytochem 36:547-550

Coombs NJ, Gough AC, Primrose JN (1999) Optimisation of DNA and RNA extraction from archival formalin-fixed tissue. Nucleic Acids Res 27:e12

Diaz-Cano SJ, Brady SP (1997) DNA extraction from formalin-fixed, paraffin-embedded tissues: protein digestion as a limiting step for retrieval of high-quality DNA. Diagn Mol Pathol 6:342-346

Emmert-Buck MR, Bonner RF, Smith PD, Chuaqui RF, Zhuang Z, Goldstein SR, Weiss RA, Liotta LA (1996) Laser capture microdissection. Science 274:998-1001

Fend F, Raffeld M (2000) Laser capture microdissection in pathology. J Clin Pathol 53:666-

672
Foss RD, Guha-Thakurta N, Conran RM, Gutman P (1994) Effects of fixative and fixation time on the extraction and polymerase chain reaction amplification of RNA from paraffin-embedded tissue. Comparison of two housekeeping gene mRNA controls. Diagn Mol Pathol 3:148-155

Frank TS, Svoboda-Newman SM, Hsi ED (1996) Comparison of methods for extracting DNA from formalin-fixed paraffin sections for nonisotopic PCR. Diagn Mol Pathol 5:220-224

Howe JR, Klimstra DS, Cordon-Cardo C (1997) DNA extraction from paraffin-embedded tissues using a salting-out procedure: a reliable method for PCR amplification of archival material. Histol Histopathol 12:595-601

Ikeda K, Monden T, Kanoh T, Tsujie M, Izawa H, Haba A, Ohnishi T, Sekimoto M, Tomita N, Shiozaki H, Monden M (1998) Extraction and analysis of diagnostically useful proteins from formalin-fixed, paraffin-embedded tissue sections. J Histochem Cytochem 46:397-403

Merkelbach S, Gehlen J, Handt S, Fuzesi L (1997) Novel enzyme immunoassay and optimized DNA extraction for the detection of polymerase-chain-reaction-amplified viral DNA from paraffin-embedded tissue. Am J Pathol 150:1537-1546

Ortiz-Pallardo ME, Ko Y, Sachinidis A, Vetter H, Fischer HP, Zhou H (2000) Detection of alpha-1-antitrypsin PiZ individuals by SSCP and DNA sequencing in formalin-fixed and paraffin-embedded tissue: a comparison with immunohistochemical analysis. J Hepatol 32:406-411

Poncin J, Mulkens J, Arends JW, de Goeij A (1999) Optimizing the APC gene mutation analysis in archival colorectal tumor tissue. Diagn Mol Pathol 8:11-19

Puchtler H, Waldrop FS, Meloan SN, Terry MS, Conner HM (1970) Methacarn (methanol-Carnoy) fixation. Practical and theoretical considerations. Histochemie 21:97-116

Scholte GH, van Doorn LJ, Quint WG, Lindeman J (1997) Polymerase chain reaction for the detection of Helicobacter pylori in formaldehyde-sublimate fixed, paraffin-embedded gastric biopsies. Diagn Mol Pathol 6:238-243

Schütze K, Lahr G (1998) Identification of expressed genes by laser-mediated manipulation of single cells. Nat Biotechnol 16:737-742

Shibutani M, Uneyama C (in press) Methacarn a fixation tool for multipurpose genetic analysis from paraffin-embedded tissues. Methods Enzymol

Shibutani M, Uneyama C. Miyazaki K, Toyoda K, Hirose M (2000) Methacarn fixation, a novel tool for analysis of gene expressions in paraffin-embedded tissue specimens. Lab Invest 80:199-208

Uneyama C, Shibutani M, Nakagawa K, Masutomi N, Hirose M (2000) Methacarn, a fixation tool for multipurpose gene expression analysis from paraffin-embedded tissue materials. Current Topics in Biochem Res 3:237-242

Whittington RJ, Reddacliff L, Marsh I, Saunders V (1999) Detection of Mycobacterium avium subsp paratuberculosis in formalin-fixed paraffin-embedded intestinal tissue by IS900 polymerase chain reaction. Aust Vet J 77:392-397

Experimental Applications (2)

Xenobiotic Metabolism

ABC transporters: a new approach to toxicogenomics

Toshihisa Ishikawa and Megumi Yoshikawa

Department of Biomolecular Engineering, Graduate School of Bioscience and Biotechnology, Tokyo Institute of Technology, 4259 Nagatsuta, Midori-ku, Yokohama 226-8501, Japan

Summary. Membrane transporters play a pivotal role in drug absorption in the intestinal tract and in drug distribution to brain and other tissues, as well as in the elimination of toxic metabolites from cells. However, at present, little is known about the toxicogenomic and pharmacogenomic significance of drug transporters. In 1992, Ishikawa first proposed a new concept of the "phase III" detoxification system by emphasizing the biological importance of ATP-dependent transporters, GS-X pumps, for elimination of glutathione S-conjugates in our body (Trends Biochem Sci 7: 463-468, 1992). Since that time, more than 40 different human ATP-binding cassette (ABC) transporter genes have been discovered, and some of them have been documented to be critically involved in transport of drugs and metabolites. This review summarizes the toxicogenomic aspects of ABC transporters and the potential mechanisms underlying their gene expression.

Key words. ABC Transporters, Phase III detoxification system, Drug metabolism, Oxidative stress, Nuclear receptors

Paradigm shift of drug discovery strategy

In the last decade of the 20th century, the development of high throughput screening (HTS) and combinatorial chemistry technologies accelerated the drug discovery process. In the 21st century, emerging genomic technologies (i.e., bioinformatics, functional genomics, and pharmacogenomics) are shifting the paradigm for drug discovery and development. However, the attrition of drug candidates in preclinical and development stages is a major problem in drug development. In about fifty percent of cases, this attrition is due to toxicity and poor pharmacokinetics (e.g., limited absorption, low plasma concentration levels, high rates of clearance). Because of increasing costs of genome-based drug discovery, pharmaceutical companies have begun to seriously re-evaluate their current strategies of drug discovery and development.

Hitherto we have been developing on a unique approach to pharmacogenomics and toxicogenomics in drug discovery research. Our strategy sprang from the realization that, in the near future, the processes of drug discovery and development will be dramatically changed by the introduction of new research technologies such as bioinformatics, functional genomics, and pharmacogenomics and their use to identify both classical drug targets (e.g., enzymes, membrane-bound receptors, and ion channels) and novel drug targets (e.g., cellular components of signal transduction, nuclear receptors, mRNA, and DNA). As a result organ-specific drug targeting and delivery will become increasingly important in ensuring the site selectivity and pharmacological activity and reducing the toxic effects of new molecular drug candidates aimed at those targets.

Importance of drug transporters

Accumulating evidence strongly suggests that drug transporters are critically involved in drug absorption, distribution and elimination, thereby affecting drug concentrations at the target site. In this context, both drug transporters and drug metabolizing enzymes can determine the pharmacokinetic profiles and overall pharmacological effects of drugs. Fig. 1 schematically illustrates the function of drug transporters in influx and efflux of drugs, as well as in elimination of toxic metabolites from cells.

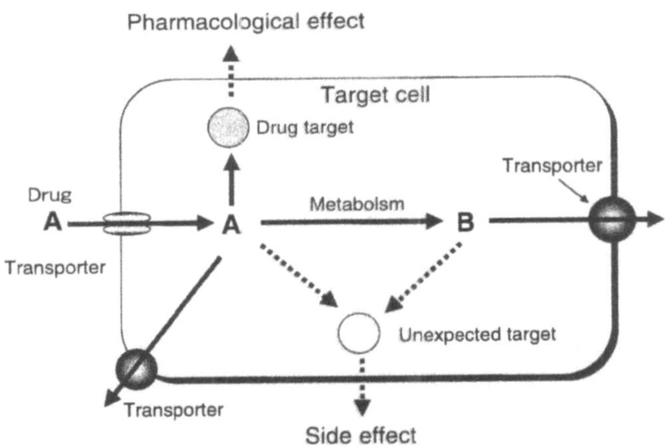

Fig. 1. Drug transporters and drug metabolizing enzymes in the drug-targeting cell.

Drug transporters expressed in epithelial cells of the small intestine and in brain capillary endothelial cells greatly affect the oral bioavailability of drugs and drug penetration into the central nervous system. Many of drug transporters belong to the ABC transporter family. This family of human ABC transporters is now known to contain about 50 members, most of whose genes have been identified and sequenced (Ishikawa et al. 2000a; Ishikawa et al. 2001).

ABCB1 (P-glycoprotein, MDR1), ABCC1 (GS-X pump, MRP1), ABCC2 (cMOAT, MRP2), and ABCG2 (BCRP, ABCP), in particular, are gaining attention for their involvement in drug absorption by the small intestine and drug penetration into the brain; they are expressed in a variety of normal cells and organs, and its modulation in these tissues can influence the activity and bioavailability of drugs. In the intestine, for instance, modulation of ABCB1 may control the degree of drug uptake after drug ingestion. At the blood-brain barrier, high ABCB1 expression levels can limit the uptake of desired drugs into the brain; conversely, low ABCB1activity can lead to abnormally increased accumulation and undesirable side effects.

ABC transporters in phase III detoxification system

Metabolism of xenobiotics including drugs is widely referred to phase I and II systems, where phase I includes oxidation of xenobiotics and phase II deals with the conjugation of phase I products (Fig. 2). The oxidative metaboism in the phase I system is mediated by cytochrome P-450 (CYP) or flavin mixed-function oxidase. Some of activated xenobiotics can interact with DNA and/or proteins in cells to cause toxic effects. In the phase II system, on the other hand, activated hydrophobic xenobiotics are converted into hydrophilic forms via conjugation reactions with glutathione, sulfate or glucuronide. This phase II metabolism is regarded as the detoxification process of xenobiotics. However, in some cases, the phase II system is a critical step in the formation of genotoxic electrophiles. Furthurmore, accumulation of the resulting metabolites in cells can lead to a decrease in the detoxification activity of the phase II system. Therefore, the phase III system must take a task to eliminate Phase II metabolites from cells. Several ABC transporters, including ABCB1, ABCC1, ABCC2 and ABCG2, are considered to be major players in the phase III detoxification system (Ishikawa 1992; Ishikawa et al. 2000b.)

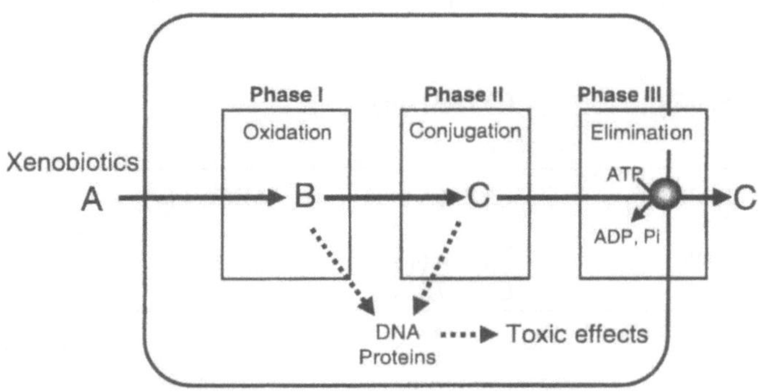

Fig. 2. Schematic illustration of phase I, II, and III detoxification systems.

Regulation of ABC transporter gene expression

Gene regulation of drug metabolizing enzymes and drug transporters is of great interest to understand the molecular mechanisms of drug response and toxic events. Recent studies have revealed that hydrophobic ligands and several nuclear receptors are involved in induction or down-regulation of cytochrome P-450 isoforms and ABC transporters (Chawla et al. 2001) (Table 1). While the gene regulation of drug metabolizing enzymes, such as cytochrome P-450, glucuronide transferases and glutathione transferases, have been well characterized, studies on the molecular mechanisms underlying the induction and down-regulation of drug transporters have just started most recently.

Table 1. Nuclear receptor-ligand interaction and expression of cytochrome P-450 (CYP) isoforms and ABC transporters.

Ligand	Nuclear receptor	CYP	ABC transporter
Xenobiotics Phenobarbital	CAR	CYP2B(+) CYP2C (+)	ABCC3 (+)
Xenobiotics Steroids	SXR/PXR	CYP3A (+) CYP2C (+)	ABCB1 (+) ABCC2 (+)
Bile acids	FXR	CYP7A1 (-) CYP8B1 (-)	ABCB11 (+)
Oxysterols	LXRα,β	CYP7A1 (+)	ABCA1 (+) ABCG1 (+), ABCG4 (+) ABCG5 (+), ABCG8 (+)
Fatty acids Fibrates	PPARα	CYP4A1 (+) CYP4A3 (+)	ABCD2 (+), ABCD3 (+) ABCB4 (+)
Fatty acids Carboprostacyclin	PPARδ	?	?
Eicosanoids Thiazolidinediones	PPARγ	CYP4AB (+)	?
Retinoic acids	RARα,β,γ	CYP26A1 (+)	ABCG4 (+)
1,25(OH)$_2$- vitamin D$_3$	VDR	CYP24 (+) CYP27B1 (-)	?

(+) up-regulation; (-) down-regulation.

Our previous studies have demonstrated that ABCC1 expression was up-regulated by heavy metals, oxidative stress and interleukin 1b (Ishikawa et al. 1996; Yamane et al. 1998; Ikegami et al. 2000). The ABCC1 gene contains core consensus sequences for ARE (antioxidant response element) and AP-1 (activator-protein-1) binding site (Fig. 3). These sequences were also found in the 5' flanking sequence of a gene encoding the active subunit of g-glutamylcysteine synthetase (g-GCS), a rate limiting enzyme of glutathione biosynthesis. It is important to note that the ABCC1 gene encodes the GS-X pump which transport a variety of glutathione conjugates. Coordinated up-regulation of ABCC1 and g-GCS genes was frequently observed in human colorectal cancers as well as in various cancer cell lines (Kuo et al 1996). Involvement of those core consensus sequences in the induction of ABCC1 and g-GCS in normal cells and tissues remains to be elucidated.

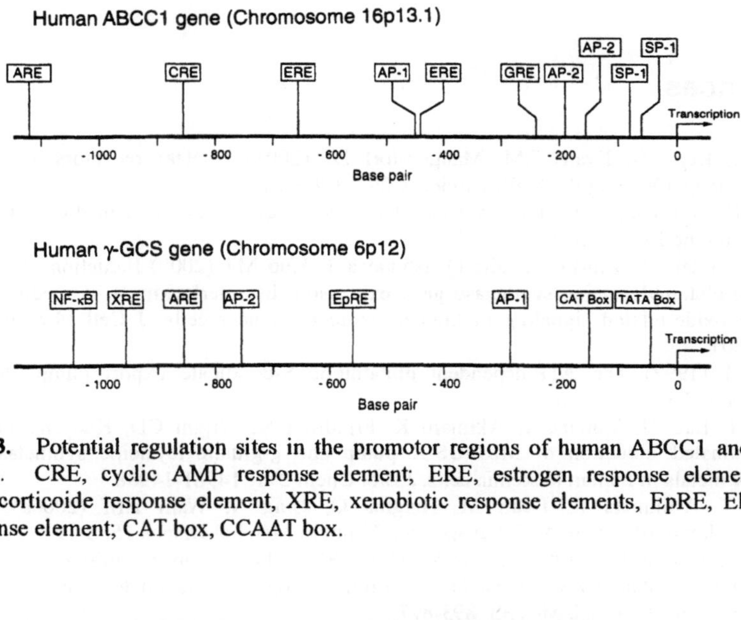

Fig. 3. Potential regulation sites in the promotor regions of human ABCC1 and g-GCS genes. CRE, cyclic AMP response element; ERE, estrogen response element; GRE glucocorticoide response element, XRE, xenobiotic response elements, EpRE, Electrphile response element; CAT box, CCAAT box.

Genetic polymorphism of ABC transporters

The effects of drug transporters on the pharmacokinetic profile of a drug depend on their expression and functionality. Indeed, the expression of drug transporters can be modulated by endogenous and exogenous factors, including drugs, themselves. It is also now known that inherited differences among individuals may also affect drug efficacy and toxicity. Such inherited differences include genetic polymorphisms in drug targets and drug-metabolizing enzymes, as well as in drug transporters. Hitherto, pharmacogenetics, the field dealing with such inherited differences and their effect on pharmacokinetics, has significantly contributed to our understanding of genetic causes underlying differences in drug metabolism (e.g., cytochrome P-450 mediated drug metabolism). In fact, recent technological advances allowing massive molecular sequencing have in turn allowed a consortium of researchers to identify single nucleotide polymorphisms (SNPs) as one possible cause of variable drug response among individuals. The SNP consortium plans to complete a high-density map of SNPs and make it openly available to the public (http://snp/cshl/org). In Japan, on the other hand, the Pharma SNP consortium (PSC) consisting of 43 pharmaceutical companies is currently investigating SNPs of drug metabolizing enzymes and drug transporters to obtain insight into the cause of individual differences in drug response (Gushima 2001; Ishikawa et al. 2001). In this context, it would become increasingly important to carefully examine the clinical significance, if any, of polymorphisms of drug transporter genes in light of both pharmacokinetic and toxicogenomic aspects.

References

Chawla A, Repa JJ, Evans RM, Mangelsdorf DJ (2001) Nuclear receptors and lipid physiology: Opening the X-files. Science 294: 1866-1870

Gushima H (2001) Japanese industry takes steps toward genomics-based medicine. Global Outsourcing Review 3: 30- 37

Ikegami Y, Tatabe S, Lin-Lee Y, Xie Q, Ishikawa T, Kuo MT (2000) Induction of MRP1 and g-glutamylcysteine synthetase gene expression by interleukin 1b is mediated by nitric oxide-related signaling in human colorectal cancer cells. J Cell Physiol 185: 293-301

Ishikawa T (1992) The ATP-dependent glutathione S-conjugate export pump. Trends Biochem Sci 17: 463-468

Ishikawa T, Bao JJ, Yamane Y, Akimaru K, Frindrich K, Wright CD, Kuo MT (1996) Coordinated induction of MRP/GS-X pump and g-glutamylcysteine synthetase by heavy metals in human leukemia cells. J Biol Chem 271: 14981-14988

Ishikawa T, Allikmets R, Dean M, Higgins C, Ling V, Wain HM (2000a) New nomenclature of human ABC transporters. Xenobio Metabol Dispos 15: 8-19

Ishikawa T, Kuo MT, Furuta K, Suzuki M (2000b) The human multidrug resistance-associated protein (MRP) gene family: From biological function to drug molecular design. Clin Chem Lab Med 38: 893-897

Ishikawa T, Yabuuchi H, Kasamatsu S, Yoshikawa M, Ikegami Y (2001) Pharmacogenomics in drug discovery and development research. Xenobio Metabol Dispos 16: 353-363

Kuo MT, Bao JJ, Curley SA, Ikeguchi M, Johnston DA, Ishikawa T (1996) Frequent coordinated overexpression of the MRP/GS-X pump and g-glutamyl-cysteine synthetase genes in human colorectal cancers. Cancer Res 56: 3642-3644

Yamane Y, Furuichi M, Song R, Yan NT, Mulcahy T, Ishikawa T, Kuo MT (1998) Expression of multidrug resistance protein/GS-X pump and g-glutamylcysteine synthetase genes is regulated by oxidative stress. J Biol Chem 271: 31075-31085

SXR and the xenobiotic response

Michelle M. Tabb, Changcheng Zhou, and Bruce Blumberg

Department of Developmental and Cell Biology, University of California, Irvine, CA 92697-2300 USA

Summary. We previously identified and characterized a nuclear receptor, SXR (steroid and xenobiotic receptor) that is the primary mediator of the xenobiotic response. SXR responds to the presence of endogenous hormones, bioactive dietary compounds, and xenobiotic chemicals by activating transcription of several P450 and drug transporter genes. The pharmacology of SXR differs considerably between rodents and humans in that a number of species-specific activators exist. Activation of SXR is thus a direct molecular assay for the potential of chemicals to exhibit divergent effects in different species. SXR-mediated differences in metabolism provide much of the mechanistic basis underlying the differential susceptibility of humans and laboratory animals to environmental chemicals. Understanding the molecular biology of SXR will enable the derivation of a commonly accepted set of principles that connect laboratory experiments, wildlife exposure data, and human risk. In turn, this will reduce the uncertainty about whether or not the underlying mechanisms of response to chemical exposure are universal, providing important new tools with which to undertake comparative studies of chemical effects between species or individuals in a population.

Keywords. Steroid and xenobiotic receptor, wild life, SXR, xenobiotic response, cytochrome P450

Introduction

When using the results of animal experiments to predict effects on humans a fundamental and often unacknowledged assumption is made. This is that uptake and metabolism of the compound in question as well as the biochemistry and endocrinology of the organism is the same between humans and the animal model. As will be discussed below, this assumption is not always valid. The mammalian xenobiotic response is mediated primarily through the activity of four families of cytochrome P450 (CYP) monooxygenases (CYP1, CYP2, CYP3, and CYP4). Of these, the CYP3 family is among the most important since it is responsible for the metabolism of more than 60% of clinically used drugs and a corresponding number of xenobiotic chemicals [24]. The CYP2B family is responsible for another 25-30% of drug and xenobiotic metabolism [37]. The ability of organisms to induce CYP enzymes in response to elevated xenobiotic levels is crucial for their survival and normal homeostasis. It has been known for some time that the induction of CYP3 genes by drugs and xenobiotics exhibits significant differences across vertebrate species [2, 9, 14, 15], thus implying that there are differences in metabolism that should be considered. Recent breakthrough studies demonstrated that a single nuclear hormone receptor termed SXR (steroid and xenobiotic receptor)[4], PXR (pregnane X receptor) [13], or PAR (pregnane activated receptor) [3] was responsible for the species-specific regulation of CYP3A genes by drugs and xenobiotic chemicals[1]. SXR is therefore a central regulator of the xenobiotic response. We review what is known about this important xenobiotic sensor and the role that toxicogenomics has to play in understanding the xenobiotic response.

SXR and species-specific xenobiotic metabolism

There is considerable evidence that wildlife and domestic animals have suffered adverse consequences from exposure to environmental chemicals . Some of these effects may result from interactions between environmental chemicals and components of the endocrine system, such as the estrogen and testosterone receptors. Increases in the prevalence of certain cancers (e.g., breast, prostate, testicular, ovarian) may be related to this "endocrine disruption". However, there is significant controversy regarding the nature of the effects (if any) induced by a particular compound and the dose at which the candidate effect is elicited as a direct result of exposure. A greater uncertainty is to what degree the data from wildlife and laboratory animal model systems can be extrapolated to measure the risk of human exposure to the same xenobiotic chemicals. In many cases, the response of animals to chemical exposure is predictive of effects on humans and therefore, appropriate toxic equivalency factors (TEF) can be formulated [31]. In other cases, the connection is more uncertain and the ability to predict human risk does not rest on sound scientific principles.

We showed that SXR is activated by a wide diversity of natural steroids (e.g., pregnanes, estranes and androstanes), dietary compounds (e.g., phytoestrogens), and xenobiotics (e.g., rifampicin, nifedipine, PCN) [4]. SXR exhibits considerable

---

[1] This receptor will be hereinafter referred to as SXR to accurately reflects its biological activity.

differences in its pharmacology among mammals [4, 12, 17], which may explain species-specific differences in xenobiotic induction of CYP3. The SXR gene is remarkably divergent across mammalian species. As would be expected for orthologous receptors, the rabbit, rodent and human receptors share ~95% amino acid sequence identity in their DNA-binding domains (Fig 1). Surprisingly, the similarity in the ligand binding domain is much less with the receptors sharing only 73-84% amino acid identity (Fig 1) rather than the ~90% typically exhibited by orthologous nuclear receptors. The putative chicken (CXR) and Xenopus (BXR) members of this family are so divergent in sequence (Fig 1) and in their activation profiles [10, 32] that it is likely they are either not orthologs or that their functions have diverged significantly during evolution.

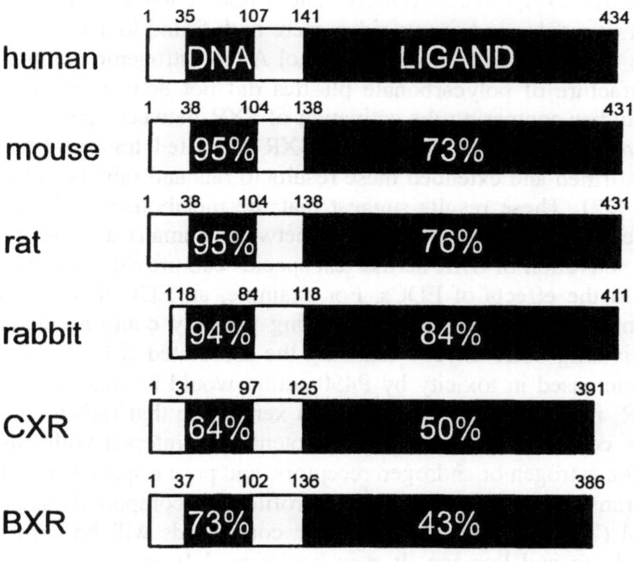

Figure 1. SXR and related genes. Percent amino acid identity is indicated among members of the NR1I12 gene family. CXR is a chicken receptor related to SXR and the bile acid receptor, CAR [10]. BXR is the Xenopus laevis benzoate 'X' receptor that was used to originally identify SXR [4].

Differences in amino acid sequence among the mammalian receptors are responsible for species-specific induction of CYP3 by drugs and xenobiotics [17, 34]. One general observation is that there are significant differences in the xenobiotic response between humans and rodents and that these differences are completely explained by the pharmacology of SXR [4, 12, 13, 17, 18, 36, 37]. For example, the antibiotic rifampicin, the anti-diabetic drug troglitazone and the cholesterol-reducing drug SR12813 were found to be effective activators of both human and rabbit SXR, but had little activity on mouse or rat SXR [12]. In contrast, pregnenolone 16α-carbonitrile was a more potent activator of rat and mouse SXR than of human or rabbit SXR [12, 19, 27, 28] (Table 1). Corresponding induction of CYP3 SXR target genes in humans, rats, and rabbits was also observed in primary hepatocytes [27].

There is increasing interest in the mechanism by which suspected endocrine disrupting compounds (EDCs) affect humans and other model organisms. This interest, together with the discovery of SXR, has prompted new studies examining the ability of EDCs to stimulate SXR-mediated transcription and CYP3 induction in an effort to determine whether there are species-specific differences in EDC metabolism. The plasticizer diethylhexylphthalate (DEHP) and 4-nonylphenol, an alkylphenol used as a component in pesticides and spermicides, were both found to activate mouse SXR and induce mouse CYP3A1. In contrast, bisphenol A, an estrogenic compound widely used in the manufacture of polycarbonate plastics did not activate mouse SXR [19, 27]. Subsequent studies comparing the activation of SXR between species showed that bisphenol A activated human but not mouse SXR mediated transcription [28]. Our laboratory has confirmed and extended these results to rats and rabbits and a variety of other compounds [27]. These results suggest that the metabolism and, by inference, the biological effects of certain EDCs differs between humans and other mammals. We conclude that activation of SXR across test species can provide an important assay for susceptibility to the effects of EDCs. For example, an EDC that is detoxified or metabolized by an SXR-dependent P450 signaling pathway could accelerate its own breakdown by activating SXR thereby reducing the associated risk. A compound that is "activated" or increased in toxicity by P450 action would be much more toxic if it also activates SXR, increasing the risk. Lastly, a xenobiotic that fails to activate SXR in a given species could then have a greater potential to interact with other nuclear receptors such as the estrogen or androgen receptors, and pose a species-specific risk.

The similar trans-species SXR activation profiles for compounds such as DEHP and 4-nonylphenol (Table 1) suggests that these compounds will be correspondingly metabolized in rodents and humans. It may be inferred from the differences in the pharmacology of human and rodent SXRs that there are classes of chemicals (e.g., phytoestrogens and bisphenol A) for which metabolism differs between humans and rodents. Therefore, rodents may not be an appropriate model system for studying these compounds. As a general principle, one should be cautious about extrapolating the results of animal testing to human risk assessment for chemicals that differentially activate SXR. Indeed, it may be necessary to separately validate animal models for each compound in question.

An important tool for the study of xenobiotic metabolism was the development of the so-called "humanized" mouse [36]. This animal is deficient in the mouse SXR gene while expressing a human SXR transgene in the liver. This model demonstrates convincingly that SXR is the key regulator of CYP3A expression. Moreover, the selective activation of target genes in response to species-specific activators was shown to reside in the ligand binding domain of the receptor, rather than in the DNA-binding domain or target DNA-binding elements [36]. This humanized mouse should prove a powerful in vivo system to predict human xenobiotic metabolism in an animal model.

Table 1. Xenobiotic activation of SXR from four mammalian species.

Ligand	*Human*	*Mouse*	*Rat*	*Rabbit*
Rifampicin	+	-	-	+
PCN	-	+	+	+
RU486	+	+	+	+
dexamethasone	-	+	+	+
trans-nonachlor	+	+	+	+
troglitazone	+	-	-	+
SR12813	+	-	-	+
Genistein	+	-	-	+
Coumestrol	+	-	-	+
Bisphenol A	+	-	-	+
4-nonylphenol	+	+	+	-
Phthalic acid	-	-	-	-
DEHP	+	+	+	+

Data is summarized from Blumberg et al. [4] Jones et al. [12] Takeshita et al. [28] and Tabb et al.[27].

Natural allelic variants of SXR

Human SXR displays a broad specificity for a variety of drugs and is a primary regulator of CYP3A4. The levels of CYP3 enzymes show considerable sexual dimorphisms and variation in levels and function among individuals in the population [8]. Variation in CYP3A expression may lead to important differences in drug metabolism, leading to clinically significant differences in drug toxicities and response. It may also influence the circulating levels of estrogens and the risk of breast cancer [16]. The molecular underpinning of the variations in CYP3A4 expression are unknown at present but it is likely that SXR plays a key role in this process. Approximately 90% of the inter-individual variability in hepatic CYP3A4 activity is genetically determined and several CYP3A4 variants have been reported [21]. However, the reported allelic frequencies and the functional data demonstrate only a limited role of these variants in CYP3A4 expression and activity [6, 33]. Single nucleotide polymorphisms (SNPs) in SXR may be a major contributor to CYP3A4 expression and activity. Since SXR acts as a xenobiotic 'sensor' to mediate the physiological response of multiple drug metabolism genes, identification of functional polymorphisms in SXR might explain the variable induction of CYP3A4 and other drug metabolizing enzymes in response to SXR ligands [5]. It has been reported that different inbred mouse lines differ substantially in their sensitivity to estrogen treatment [25]. The observed differential sensitivity may result from differences in SXR among these strains although this remains to be demonstrated.

The SXR gene consists of nine exons and spans approximately 35 kb in chromosome 13q11-13. Recently, SXR variants were investigated in two ethnic groups: Caucasians and Africans. More than 40 SNPs were identified including seven in the coding region that are non-synonymous, creating new SXR alleles [11, 38]. Three of the seven variants were located N-terminal to the DNA binding domain (E18K, P27S, and G36R) and have no significant effects on DNA-binding or transactivation compared with wild-type SXR. One rare variant (R122Q) is located in the third helix of the SXR DBD. This mutant shows significantly decreased affinity for DNA binding and attenuated transcriptional activity. The other three variants are within the LBD of SXR (D163G and A370T) or close to the LBD (V140M), and show some alterations in the activation properties of SXR [11]. The A370T and V140M variants show 1.5 –2 fold enhancement in the basal expression of a CYP3A4 promoter reporter gene but lack any significant effect on transcriptional activation. In contrast, the D163G variant exhibits lower basal activity and an eight-fold higher induction by rifampicin than wild-type SXR [11].

It may be concluded that variation in SXR plays a role in the inter-individual variability of CYP3A4 expression and drug response. In addition, SXR polymorphisms could also influence individual predisposition to tumors caused by environmental carcinogens, including liver and lung cancer [7, 22]. Recently, SXR was also shown to be a lithocholic acid sensor that controls the biosynthesis and metabolism of bile acids [35]. Thus, SXR variants may be associated with physiological and pathophysiological changes in steroid, cholesterol or bile acid levels as well as the xenobiotic response.

SXR mediates xenobiotic metabolism and efflux

Many pharmacological agents are transported out of cells by intestinal P-glycoprotein (MDR1) and inactivated by liver cytochrome P450 enzymes. Both P-glycoprotein and P450 enzymes possess broad substrate specificity and the genes encoding these proteins share the ability to be upregulated by the nuclear receptor SXR [26]. Its expression in hepatocytes, cells lining the intestine and in proximal renal tubules [29, 30] suggests that P-glycoprotein plays a physiological role as a defense mechanism against potential toxic substances encountered in the diet and through environmental exposure. Both SXR and MDR1 are co-expressed in a variety of tissues including liver, kidney and placenta allowing a coordinate regulation of metabolism and excretion via xenobiotic activation of SXR.

Expression of MDR1 can be problematic in designing effective strategies for cancer treatment and other forms of chemotherapy. Identification of SXR as a key regulator of MDR1 suggests that chemotherapies which do not activate SXR and thereby induce drug clearance might be more effective. This is illustrated by the efficacy of taxol versus taxotere as anti-neoplastic agents. The superior pharmacokinetic properties of taxotere are due to its inability to activate SXR and induce SXR target genes involved in drug clearance [26]. Just as genetic variability plays a role in the pharmacology of SXR, germline polymorphisms have been found in MDR1, and these could modulate response to xenobiotics and predict the success of chemotherapies in different individuals [23]. Likewise, this same genetic variability could determine how well an individual is able to clear cells of an assault by harmful xenobiotics or EDCs.

Placental expression of SXR and its target genes may indicate a regulation of fetal exposure to endogenous steroids and xenobiotics *in utero*. Cytochrome P450s and P-glycoprotein expression have been detected in placenta [1, 20]. The unique transient high levels of steroid hormone that a fetus is exposed to during pregnancy induce SXR and CYP3A target gene expression in perinatal mice implying a role for SXR in regulation of fetal steroid hormone levels during pregnancy. Likewise, fetal exposure to, or protection from the effects of harmful xenobiotics or EDCs may be determined by regulation via SXR in the placenta. Whether or not an EDC is able to activate or block the action of human SXR could determine downstream effects on fetal development including malformations and endocrine disruption.

Conclusion and Prospects

Understanding how the xenobiotic response differs between individual humans, different animal species or even among different laboratory strains of the same species is essential to developing high quality models and characterizations of risk from chemical exposure. The recent identification of the nuclear receptor SXR provides a key that can be used to greatly expand our understanding of the pathways that control and modulate the response to toxic environmental chemicals in the body. It is now clear that there are many commonalities but also specific differences in the way organisms respond to chemical exposure. A significant difficulty in deriving a commonly accepted set of principles that connect laboratory experiments, wildlife exposure data and human risk is the uncertainty about whether the underlying mechanisms of response to chemical exposure are universal or different. SXR shows strong similarities as well as important differences in its response to natural and xenobiotic chemicals across mammalian species. SXR may be used as an assay to identify chemicals that are metabolized differently between humans and rodents and even between humans. This provides an important new tool with which to undertake comparative studies of chemical effects between individuals in a population, whether it is inbred laboratory animal strains or different human ethnic groups.

Toxicogenomics will be increasingly important in furthering our understanding of the xenobiotic response. It will be necessary to identify the full spectrum of SXR-responsive genes and to determine which of these are responsible for individual variations in xenobiotic and drug metabolism. High throughput screening and profiling techniques will generate fingerprints of exposure to particular xenobiotics and enable the rapid detection of individual differences in gene expression. This will facilitate a molecular dissection of the xenobiotic response and lead to the development of improved tools for risk assessment.

Acknowledgements:

We thank R. Kaigh for critically reading the manuscript. Work in the author's laboratory was supported by a grant from the National Cancer Institute (CA-87222) and gifts from Eisai Co., Ltd (Japan) and the Minamata Environmental Research and Development Center Co., Ltd. (Japan).

References

1. Ambudkar SV, Dey S, Hrycyna CA, Ramachandra M, Pastan I, Gottesman MM (1999) Biochemical, cellular, and pharmacological aspects of the multidrug transporter. Annu Rev Pharmacol Toxicol 39:361-98

2. Barwick JL, Quattrochi LC, Mills AS, Potenza C, Tukey RH, Guzelian PS (1996) Trans-species gene transfer for analysis of glucocorticoid-inducible transcriptional activation of transiently expressed human CYP3A4 and rabbit CYP3A6 in primary cultures of adult rat and rabbit hepatocytes. Mol Pharmacol 50(1): 10-6
3. Bertilsson, G., J. Heidrich, K. Svensson, M. Asman, L. Jendeberg, M. Sydow-Backman, R. Ohlsson, H. Postlind, P. Blomquist, and A. Berkenstam. 1998. Identification of a human nuclear receptor defines a new signaling pathway for CYP3A induction. Proc Natl Acad Sci U S A 95(21): 12208-13.
4. Blumberg, B., W. Sabbagh, Jr., H. Juguilon, J. Bolado, Jr., C. M. van Meter, E. S. Ong, and R. M. Evans. 1998. SXR, a novel steroid and xenobiotic-sensing nuclear receptor. Genes Dev 12(20): 3195-205.
5. Eichelbaum, M., and O. Burk. 2001. CYP3A genetics in drug metabolism. Nat Med 7(3): 285-7.
6. Eiselt, R., T. L. Domanski, A. Zibat, R. Mueller, E. Presecan-Siedel, E. Hustert, U. M. Zanger, J. Brockmoller, H. P. Klenk, U. A. Meyer, K. K. Khan, Y. A. He, J. R. Halpert, and L. Wojnowski. 2001. Identification and functional characterization of eight CYP3A4 protein variants. Pharmacogenetics 11(5): 447-58.
7. Forrester, L. M., G. E. Neal, D. J. Judah, M. J. Glancey, and C. R. Wolf. 1990. Evidence for involvement of multiple forms of cytochrome P-450 in aflatoxin B1 metabolism in human liver. Proc Natl Acad Sci U S A 87(21): 8306-10.
8. Gonzalez, F. J. 1992. Human cytochromes P450: problems and prospects. Trends Pharmacol. Sci. 13: 346-352.
9. Guzelian, P. S. 1988. Regulation of the glucocorticoid-inducible cytochromes P450, p. 148-155. *In* J. O. Miners, D. J. Birkett, R. Drew, and M. McManus (ed.), Microsomes and Drug Oxidations. Taylor and Francis, London, U.K.
10. Handschin, C., M. Podvinec, and U. A. Meyer. 2000. CXR, a chicken xenobiotic-sensing orphan nuclear receptor, is related to both mammalian pregnane X receptor (PXR) and constitutive androstane receptor (CAR). Proc Natl Acad Sci U S A 97(20): 10769-74.
11. Hustert, E., A. Zibat, E. Presecan-Siedel, R. Eiselt, R. Mueller, C. Fuss, I. Brehm, U. Brinkmann, M. Eichelbaum, L. Wojnowski, and O. Burk. 2001. Natural protein variants of pregnane X receptor with altered transactivation activity toward CYP3A4. Drug Metab Dispos 29(11): 1454-9.
12. Jones, S. A., L. B. Moore, J. L. Shenk, G. B. Wisely, G. A. Hamilton, D. D. McKee, N. C. Tomkinson, E. L. LeCluyse, M. H. Lambert, T. M. Willson, S. A. Kliewer, and J. T. Moore. 2000. The pregnane X receptor: a promiscuous xenobiotic receptor that has diverged during evolution. Mol Endocrinol 14(1): 27-39.
13. Kliewer, S. A., J. T. Moore, L. Wade, J. L. Staudinger, M. A. Watson, S. A. Jones, D. D. McKee, B. B. Oliver, T. M. Willson, R. H. Zetterstrom, T. Perlmann, and J. M. Lehmann. 1998. An orphan nuclear receptor activated by pregnanes defines a novel steroid signaling pathway. Cell 92(1): 73-82.
14. Kocarek, T. A., E. G. Schuetz, and P. S. Guzelian. 1993. Regulation of phenobarbital-inducible cytochrome P450 2B1/2 mRNA by lovastatin and oxysterols in primary cultures of adult rat hepatocytes. Toxicol Appl Pharmacol 120(2): 298-307.

15. Kocarek, T. A., E. G. Schuetz, S. C. Strom, R. A. Fisher, and P. S. Guzelian. 1995. Comparative analysis of cytochrome P4503A induction in primary cultures of rat, rabbit, and human hepatocytes. Drug Metab Dispos 23(3): 415-21.
16. Kuehl, P., J. Zhang, Y. Lin, J. Lamba, M. Assem, J. Schuetz, P. B. Watkins, A. Daly, S. A. Wrighton, S. D. Hall, P. Maurel, M. Relling, C. Brimer, K. Yasuda, R. Venkataramanan, S. Strom, K. Thummel, M. S. Boguski, and E. Schuetz. 2001. Sequence diversity in CYP3A promoters and characterization of the genetic basis of polymorphic CYP3A5 expression. Nat Genet 27(4): 383-91.
17. LeCluyse, E. L. 2001. Pregnane X receptor: molecular basis for species differences in CYP3A induction by xenobiotics. Chem Biol Interact 134(3): 283-9.
18. Lehmann, J. M., D. D. McKee, M. A. Watson, T. M. Willson, J. T. Moore, and S. A. Kliewer. 1998. The human orphan nuclear receptor PXR is activated by compounds that regulate CYP3A4 gene expression and cause drug interactions. J Clin Invest 102(5): 1016-23.
19. Masuyama, H., Y. Hiramatsu, M. Kunitomi, T. Kudo, and P. N. MacDonald. 2000. Endocrine disrupting chemicals, phthalic acid and nonylphenol, activate Pregnane X receptor-mediated transcription. Mol Endocrinol 14(3): 421-8.
20. Masuyama, H., Y. Hiramatsu, Y. Mizutani, H. Inoshita, and T. Kudo. 2001. The expression of pregnane X receptor and its target gene, cytochrome P450 3A1, in perinatal mouse. Mol Cell Endocrinol 172(1-2): 47-56.
21. Ozdemir, V., W. Kalowa, B. K. Tang, A. D. Paterson, S. E. Walker, L. Endrenyi, and A. D. Kashuba. 2000. Evaluation of the genetic component of variability in CYP3A4 activity: a repeated drug administration method. Pharmacogenetics 10(5): 373-88.
22. Paolini, M., G. Cantelli-Forti, P. Perocco, G. F. Pedulli, S. Z. Abdel-Rahman, and M. S. Legator. 1999. Co-carcinogenic effect of beta-carotene. Nature 398(6730): 760-1.
23. Potocnik, U., M. R. Glavac, R. Golouh, and D. Glavac. 2001. The role of P-glycoprotein (MDR1) polymorphisms and mutations in colorectal cancer. Pflugers Arch 442(6): R182-3.
24. Rendic, S., and F. J. Di Carlo. 1997. Human cytochrome P450 enzymes: a status report summarizing their reactions, substrates, inducers, and inhibitors. Drug Metab Rev 29(1-2): 413-580.
25. Spearow, J. L., P. Doemeny, R. Sera, R. Leffler, and M. Barkley. 1999. Genetic variation in susceptibility to endocrine disruption by estrogen in mice. Science 285(5431): 1259-61.
26. Synold, T. W., I. Dussault, and B. M. Forman. 2001. The orphan nuclear receptor SXR coordinately regulates drug metabolism and efflux. Nat Med 7(5): 584-90.
27. Tabb, M. M., C. Zhou, D. Hemmati, and B. Blumberg. 2002. manuscript in preparation.
28. Takeshita, A., N. Koibuchi, J. Oka, M. Taguchi, Y. Shishiba, and Y. Ozawa. 2001. Bisphenol-A, an environmental estrogen, activates the human orphan nuclear receptor, steroid and xenobiotic receptor-mediated transcription. Eur J Endocrinol 145(4): 513-7.
29. Thiebaut, F., T. Tsuruo, H. Hamada, M. M. Gottesman, I. Pastan, and M. C. Willingham. 1987. Cellular localization of the multidrug-resistance gene product P- glycoprotein in normal human tissues. Proc Natl Acad Sci U S A 84(21): 7735-8.
30. Thiebaut, F., T. Tsuruo, H. Hamada, M. M. Gottesman, I. Pastan, and M. C. Willingham. 1989. Immunohistochemical localization in normal tissues of different epitopes in the

multidrug transport protein P170: evidence for localization in brain capillaries and crossreactivity of one antibody with a muscle protein. J Histochem Cytochem 37(2): 159-64.

31. Van den Berg, M., L. Birnbaum, A. T. Bosveld, B. Brunstrom, P. Cook, M. Feeley, J. P. Giesy, A. Hanberg, R. Hasegawa, S. W. Kennedy, T. Kubiak, J. C. Larsen, F. X. van Leeuwen, A. K. Liem, C. Nolt, R. E. Peterson, L. Poellinger, S. Safe, D. Schrenk, D. Tillitt, M. Tysklind, M. Younes, F. Waern, and T. Zacharewski. 1998. Toxic equivalency factors (TEFs) for PCBs, PCDDs, PCDFs for humans and wildlife. Environ Health Perspect 106(12): 775-92.
32. Venkatesan, R., J. Cao, F. Grun, and B. Blumberg. 2002. The Xenopus orphan nuclear receptor BXR is not a xenobiotic sensor. manuscript in preparation.
33. Wandel, C., J. S. Witte, J. M. Hall, C. M. Stein, A. J. Wood, and G. R. Wilkinson. 2000. CYP3A activity in African American and European American men: population differences and functional effect of the CYP3A4*1B5'- promoter region polymorphism. Clin Pharmacol Ther 68(1): 82-91.
34. Watkins, R. E., G. B. Wisely, L. B. Moore, J. L. Collins, M. H. Lambert, S. P. Williams, T. M. Willson, S. A. Kliewer, and M. R. Redinbo. 2001. The human nuclear xenobiotic receptor PXR: structural determinants of directed promiscuity. Science 292(5525): 2329-33.
35. Willson, T. M., S. A. Jones, J. T. Moore, and S. A. Kliewer. 2001. Chemical genomics: functional analysis of orphan nuclear receptors in the regulation of bile acid metabolism. Med Res Rev 21(6): 513-22.
36. Xie, W., J. L. Barwick, M. Downes, B. Blumberg, C. M. Simon, M. C. Nelson, B. A. Neuschwander-Tetri, E. M. Brunt, P. S. Guzelian, and R. M. Evans. 2000. Humanized xenobiotic response in mice expressing nuclear receptor SXR. Nature 406(6794): 435-9.
37. Xie, W., and R. M. Evans. 2001. Orphan nuclear receptors: the exotics of xenobiotics. J Biol Chem 276(41): 37739-42.
38. Zhang, J., P. Kuehl, E. D. Green, J. W. Touchman, P. B. Watkins, A. Daly, S. D. Hall, P. Maurel, M. Relling, C. Brimer, K. Yasuda, S. A. Wrighton, M. Hancock, R. B. Kim, S. Strom, K. Thummel, C. G. Russell, J. R. Hudson, Jr., E. G. Schuetz, and M. S. Boguski. 2001. The human pregnane X receptor: genomic structure and identification and functional characterization of natural allelic variants. Pharmacogenetics 11(7): 555-72.

Experimental Applications (3)

Endocrine Disruptor Research

C. elegans as a tool for environmental toxicology

Koji Arizono,[1,5] Kazuhiro Ura,[1,5] Nobuaki Tominaga,[2] Toshinori Kai,[1]
Yuji Kohara,[3] and Taisen Iguchi[4,5]

[1] Faculty of Environmental and Symbiotic Sciences, Prefectural University of Kumamoto, Japan
[2] Ariake National College of Technology, Omuta, Japan
[3] Center for Genetic Resonance Information, National Institute of Genetics, Mishima, Japan
[4] National Institute of Basic Biology, Okazaki National Research Institutes, Okazaki, Japan
[5] CREST, JST

Summary. We describe the utility of the free-living soil nematode *Caenorhabditis elegans* (*C. elegans*) as a model organism for toxicity tests. Dose response data were observed on larvae *C. elegans* exposed to xenobiotic chemicals, $CdCl_2$ and Bisphenol-A (BPA), under liquid toxicity test system. The end points measured were 5, 8, 24 hr mortality. The LD_{50} value was low level or the same level in *C. elegans*, although a longer duration was needed in yeast toxicity test. For mortality, $CdCl_2$ was more toxic than BPA. In addition to measurement of mortality after chemical exposure, we analyzed the effects of xenobiotic chemicals at the molecular level in *C. elegans*. In order to identify genes that are differentially expressed as a consequence of toxic stress due to $CdCl_2$ or BPA, we used the cDNA microarray system to compare mRNA expression patterns in *C. elegans*. Using cDNA microarray technique, we found that a large number of target genes positively and negatively regulated by each chemicals in *C. elegans*. Chemicals affected gene expression for a diverse range of cellular functions and the differential gene expression pattern was observed between different chemical exposures. These data suggest that *C. elegans* is a suitable test organism for chemical bioassay of environmental xenobiotic chemicals.

Key words. Nematoda, *Caenorhabditis elegans*, Toxicology, cDNA microarray

Introduction

During the last several decades the global environment has been polluted by a number of natural and xenobiotic compounds including a wide variety of endocrine-disrupting chemicals (EDCs). These chemical compounds affect several physiological system, development, growth, reproduction and behavior of human and wildlife. There is growing concern about a range of substances, which may cause adverse health effects such as behavioral and reproductive abnormalities. To solve these problems, it is essential to estimate exactly chemical value and their toxicities, and to elucidate the mechanisms of toxic action in human and wildlife.

To date, various monitoring systems have been established such as the in vivo test, which uses the individual organisms such as fish, water flea and mouse, and the in vitro system which uses cultured yeast cell, and the luminescence bacteria. The vivo test can estimate the overall influence of the chemical on the living organism. The in vivo test however, has major disadvantages such as long breeding and chemical exposure periods, and the issue of individual differences. The equipment and technology required are also a problem. In vitro tests on the other hands have short exposure periods and it is possible to estimate the effects of lots of chemicals at the same time. Generally, it needs that after focused in vitro toxicity assays, chemicals lead proceed in to animal tests, in vivo. In this study, we examined *C. elegans* as a model organism for toxicity tests.

C. elegans as a model for toxicology

What is a *C. elegans*

In 1974, Sydney Brenner published a landmark paper heralding *C. elegans* as a model organism for biomedical research (Brenner 1974). *C. elegans* is free-living nematode with a transparent body and natural inhabitant of the liquid phase of soil. *C. elegans* naturally exists as either a self-fertile hermaphrodite capable of producing >300 self-progeny or a male that can cross-fertilize hermaphrodites. The adult hermaphrodite is composed of 959 somatic cells but contains highly differentiated muscle, nervous, digestive, and reproductive systems, although their body length is only 1mm long. Also they have a short life cycle of 3-day at 20 ℃ (Sulston 1998).

Why use *C.elegans*

C. elegans is amenable to genetic, molecular, cell biology, and biochemical analysis. It grows easily on agar plates or in liquid medium containing *E. coli* as food, and does not require large space. The rapid life cycle allows genetic screens to be performed over multiple generations. The ability to grow *C. elegans* in liquid culture singly or in bulk is being used to screen for chemicals to identify resistance or hypersensitive mutants. In addition, *C. elegans* is a simple multicellular organism whose genome has been fully sequenced. The nematode genome is relatively small (~ 10^8 bp), and abundant information is available on the genetic and physical maps of its chromosomes (Waterston et al., 1997). High levels of evolutionally conservation between nematode and higher organisms are observed in amino acid sequence of many gene products (Consortium 1998). The development and function of the diploid organism is encoded by at least 17,000 distinct genes (Reboul et al., 2001). Moreover, high-throughput genomic analysis projects have been initiated; and the data is readily available through several databases (WormeBase; The Nematode Expression Pattern DataBase;

Sanger Center the *C. elegans* Genome Project).

Due to these advantages, considerable amounts of toxicity test have been performed using *C. elegans*. Mortality was assessed in liquid medium (Williams and Dusenbery 1990), in soil (Donkin and Dusenbery 1993), on agar plates (Williams and Dusenbery 1988), changes in movement (Anderson et al., 2001), feeding rate (Anderson et al., 2001), stress protein production (Guven et al., 1994), fecundity (Hoss et al., 1999, Anderson et al., 2001), development (Hoss et al., 1999).

The evaluation of safety test for environmental chemicals

The assessment of 50% lethality after a chemical exposure is one of traditional and popular toxicity tests. We routinely use this toxicity assessment in liquid phase using 96 well culture plates.

Nematode (N2 strain) was cultured on NGM plate with *E. coli* DH5α as a food source, and cultures at 20℃ for 2-3days. Same size (= same stage) worms (L2-L3 larvae) were selected by loading on a column of Sephadex G-25. For each of the 3 replicates, transferred 200ul of each concentration of chemical solution per 1 well in 96 well plates, and then applied about 20 larvae in each well. Worms exposed for 0, 5, 8, 24 hr at 20℃. After chemical exposure, each well was observed under dissecting microscope, and then counted living worms and dead worms. We decided that worms vanished movement was dead. Survival rate was calculated by number of living worms per all worms.

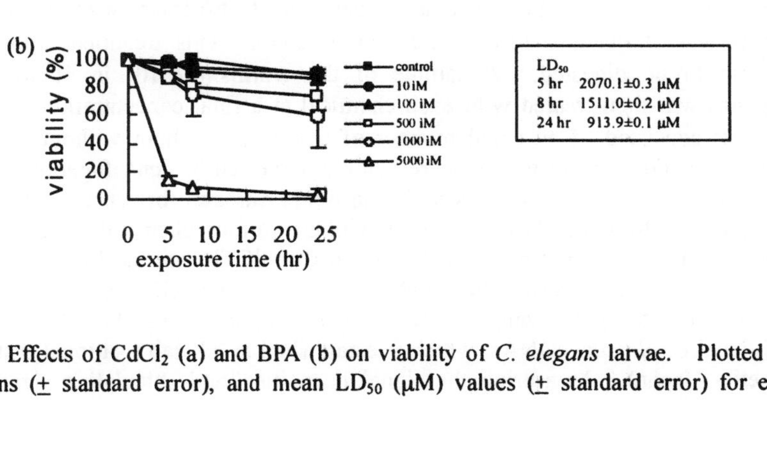

Fig. 1. Effects of $CdCl_2$ (a) and BPA (b) on viability of *C. elegans* larvae. Plotted are the means (± standard error), and mean LD_{50} (μM) values (± standard error) for each

duration time are provided.

Toxicity was examined by using $CdCl_2$ and Bisphenol-A (BPA); half the number of them at each time calculated a fatal value (LD_{50} value) this time (Fig. 1). The LD_{50} value (The chemicals were exposed for eight hours) was low level ($CdCl_2$) or the same level (BPA) as yeast (UK-Japan research cooperation on endocrine disrupting chemicals, 2001). Then, the exposure time in the toxicity of the nematode was assumed to be eight hours. The medaka toxicity test is 96 hours necessary, and the daphnia toxicity test is needed for 48 hours (UK-Japan research cooperation on endocrine disrupting chemicals, 2001). The toxicity test needs long exposure time in other animals. However, *C. elegans* was able to estimate almost equal toxicity at short exposure time for 8 hr in this study.

Recently, the harmful effects on reproductive system by EDCs in human and wildlife have been focused. It was also observed the decreasing of fecundity rate was observed by several steroid hormones and synthesized steroids in *C. elegans* using the agar plate assay (Tominaga et al., in submitted). As for the toxicity evaluation system, which used *C. elegans*, utility and handiness, was suggested.

Detection of chemical responsive gene by cDNA microarray

To date, various methods have been used for detecting gene expression levels and identifying differentially expressed genes to clarify the mechanism of chemical actions in vivo or in vitro experiments. Recently, the DNA microarray system have been established, and used to study different gene expression profiles of different physiological state of *C. elegans* (Hill et al., 2000, Reinke et al., 2000, Jiang et al., 2001). This system is a powerful tool to study chemical effects on cellular metabolism and gene regulation on a genomic scale in animals (Welsh et al., 2001). In the current study, we adopted a cDNA microarry system to detect differentially expressed gene by chemical exposure in *C. elegans*.

The survival of *C. elegans* in the presence of $CdCl_2$ or BPA, within the exposure time range and serial dilutions are shown in Figure 1. In 10 μM $CdCl_2$ and BPA, the survival rate after 5 hr was > 90% and this concentration and time were chosen as the experimental parameters for the cDNA microarray analysis.

Following cleaning of worm cultures and prior to treatment with chemicals the worms was allowed to recover 30 min at 20℃. This treatment allowed gut bacteria to be digested. A sample of the worms ($-2x10^6$ in 100ml culture medium) was then treated with each chemical at a final concentration of 10 μM. Controls consisted of an equal number of worms in medium without chemicals. After incubation for 5hr worms were quickly harvested by centrifugation at x780g, and then samples were frozen as pellets in liquid nitrogen for subsequent analyses. The mRNAs from the larvae with or without various chemical exposures were isolated. Different fluorophores (Cy3- and Cy5-dCTP) are used to label cDNAs from control and experimental mRNAs. The labeled cDNAs are then mixed together prior to hybridization to cDNA microarray including about 8000 kinds of individual *C. elegans* cDNAs, which is established in YK lab (National Institute of Genetics, Japan). Measuring the signal intensities, and calculating signal ratios determine relative amounts of a particular gene transcript in the two samples.

Table 1. Effects of $CdCl_2$ and BPA on differentially expressed gene number using cDNA microarray analysis. The criteria for positive- and negative-regulated gene were that the ratio of expression level in the chemical exposure sample to the expression level in the control sample was greater than 2.0, 0.5, respectively.

	$CdCl_2$	BPA
Total analyzed gene	8,348	8,348
Positive-regulated gene	25	11
Negative-regulated gene	6	38
Unchanged gene	4,685	4,986
Non-detected gene	3,632	3,313

Using cDNA microarray analysis, we found that a large number of target genes showed positive and negative expression when *C. elegans* was exposed to $CdCl_2$ and BPA (Table 1). Chemicals affected gene expression for a diverse range of cellular functions, including protein synthesis, protein modification, small molecule transport, and signal transduction etc., and the differential gene expression pattern was observed. The ability of cadmium to induce metallothionein gene (MT) expression in a variety of species has been documented (Hamer 1986). In this study, although MT and MT-linking gene were induced by cadmium, these genes were not induced by BPA. On the other hand, expression of a several gene associated lipid, fatty acid and sterol metabolism were regulated by BPA. Cadmium however has no effect on these gene expressions. This data suggests that there are different physiological cascades in chemical response mechanisms of *C. elegans*. Moreover, our results indicate that the cDNA microarray analysis is useful for identifying novel target genes of the various chemicals due to construct a "Toxico-DNA array".

References

Anderson GL, Boyd WA, Williams PL (2001) Assessment of sublethal endpoints for toxicity testing with the nematode Caenorhabditis elegans. Environ Toxicol Chem 20:833-838

Brenner S (1974) The genetics of Caenorhabditis elegans. Genetics 77: 71-94

Consortium (The *C. elegans* Sequencing Consortium) (1998) Genome sequence of the nematode *C. elegans*: a platform for investigating biology. Science 282:2012-2018

Donkin SG, Dusenbery DB (1993) Soil toxicity test using the nematode Caenorhabditis and an effective method of recovery. Arc Environ Contam Toxicol 25:145-151

Guven K, de Pomerai DI (1995) Differential expression of hsp70 proteins in response to heat and cadmium in Caenorhabditis elegans. J Thermal Biol 20:355-363

Hamer D (1986) Metallothionein. Annu Rev Biochem 55:913-951

Hill AA, Hunter CP, Tsung BT, Tucker-Kellogg G, Brown EL (2000) Genomic analysis of gene expression in *C. elegans*. Science 290:809-812

Hoss S, Haitzer M, Traunsurger W, Steinberg CEW (1999) Growth and fertility of Caenorhabditis elegans (nematoda) in unpoluted freshwater sediments: response to particle size disturibution and oraganic content. Environ Toxicol Chem 18:2921-2925

Jiang M, Ryu J, Kiraly M, Duke K, Reinke V, Kim SK (2001) Genome-wide analysis of developmental and sex-regulated gene expression profiles in Caenorhabditis elegans. Proc Natl Acad Sci USA 98:218-223

Reboul J, Vaglio P, Tzellas N, Thierry-Mieg N, Moore T, Jackson C, Shin-I T, Kohara Y, Thierry-Mieg D, Thierry-Mieg J, Lee H, Hitti J, Doucette-Stamm L, Hartley JL, Temple GF, Brasch MA, Vandenhaute J, Lamesch PE, Hill DE, Vidal M (2001) Open-reading-frame sequence tags (OSTs) support the existence of at least 17,300 genes in *C. elegans*. Nature Genetics 27:332-336

Reinke V, Smith HE, Nance J, Wang J, Doren CV, Begly R, Jones SJM, Davis EB, Schere S, Ward S, Kim SK (2000) A global profile of germline gene expression in *C. elegans*. Molecular Cell 6:605-616

Sulston J (1988) in The Nematode Caenorhabditis elegans (Wood, WB ed) pp. 123-155, Cold Spring Harbor laboratory, Cold Spring Harbor, NY

Waterston RH, Sulston JE, Coulson AR (1997) in *C. elegans* II (Riddle DL, Blumenthal T, Meyer BJ, Priess JR eds) pp. 23-46, Cold Spring Harbor laboratory, Cold Spring Harbor, NY

Welsh JB, Zarrinkar PP, Sapinoso LM, Kern SG, Behling CA, Monk BJ, Lockhart DJ, Burger RA, Hampton GM (2001) Analysis of gene expression profiles in normal and neoplastic ovarian tissue samples identifies candidate molecular markers of epithelial ovarian cancer. Proc Natl Acad Sci USA 98:1176-1181

Williams PL, Dusenbery DB (1988) Using Caenorhabditis elegans to predict mammalian acute lethality to metalic salts. Toxicol Ind Health 4:469-478

Williams PL, Dusenbery DB (1990) Aquatic toxicity testing using the nematode Caenorhabditis elegans. Environ Toxicol Chem 9:1285-1290

Development of the test method for detection of endocrine-disrupting activity using DNA microarrays

Akihiro Kondo[1] and Ken Takeda[2]

[1]Biotechnology Research Laboratory, Takara Shuzo Co., Ltd., Seta 3-4-1, Otsu, Shiga 520-2193, Japan.
[2]Faculty of Pharmaceutical Sciences, Tokyo University of Science, Ichigaya, Shinjuku-ku, Tokyo 162-8601, Japan.

Summary. Using recent developments of DNA microarray technology, the expression and/or suppression of tens of thousands of genes can be measured simultaneously. DNA microarray technology has a wide applicability to the fields of the research on the effects of endocrine disrupting chemicals (EDCs). Here, we present the application of this technology to studies of developing and validating methods for risk assessment of EDCs. We have examined gene expression regulation by the test chemicals on steroid-hormone-sensitive human and mouse cell lines. In the case of human cell lines, a microarray comprising about 8,400 human genes was used for the test. A statistically significant change in expression was observed for about 1,000 genes in the microarray. In the case of mice, about 1,500 of approximately 9,000 genes were identified. These selected genes can be called "expression regulated genes associated with EDC effects". DNA microarrays could provide not only a method to quickly categorize chemicals and assign a mode of adverse effects but also more sensitive end points to be addressed by gene expression regulation.

Key words. DNA microarray, endocrine disrupting chemicals, low dose effects, estrogen

Recently, an increased awareness of endocrine disrupting chemicals (EDCs) and their potential to affect wildlife as well as humans have produced a demand for practical detection methods to identify endocrine disrupting activity of EDCs. Some reports in the last few years describing apparent increase in developmental or reproductive abnormalities in wildlife or humans have resulted in a worldwide intensification of research efforts to characterize EDCs (Colborn et al. 1993, Harrison et al. 1997). Many of these chemicals mimic or block the activity of natural estrogens by specifically binding to the estrogen receptor (ER), resulting in the transcriptional control of a variety of genes in target cells. Several test methods have been developed such as estrogen receptor binding assay (Bolger et al. 1998), E-screen assay (Soto et al. 1995), reporter gene assay (Metzger et al. 1988, Nishikawa et al. 1999), and uterotrophic assay (Gordon et al. 1986, Reel et al.

1996). Those test methods are focused on measurement of the strength of estrogenic activity. Among EDCs found to cause endocrine disruption in wildlife and humans there are estrogen mimics. However, what we want to know is EDCs have not only estrogenic activity but also endocrine disrupting activity. Moreover, we want to have reasonable answers to the questions "How to make sure of the mechanism of EDCs' function?" and "How to define the endocrine disruption effects of EDCs on wildlife and humans?" In other words, the challenge in resolving the endocrine disruption issue is not enough to substantiate whether EDCs result in adverse effects to the endocrine systems of wildlife and humans.

The effects, or "toxicity", caused by EDCs are not traditionally examined thoroughly in conventional toxicological studies. This is because the effects of EDCs are mainly mediated through endocrine receptors that are known to involve new paradigms of toxicity, such as low dose effects, unusual dose-response relationships, and possible direct and indirect effects on fetuses. "Low dose effects" refers to biological changes that occur at doses lower than those typically used in standard toxicological testing for the evaluation of reproductive and developmental toxicity. Some recent observations regarding induced endocrine disruption in animals has stemmed from investigation of the possibility that exposure to low levels of EDCs can result in adverse effects (Nagel et al. 1997, vom Saal et al. 1998, Howdeshell et al. 1999). However, other researchers have reported failure to repeat these observations. When observations are found to be difficult to repeat, the original observation of low dose effects is not invalidated. To clarify uncertainties and better characterize low dose effects, we recommended using genetic approaches to determine mechanisms of action and to characterize the dose-response relationship. In other words, traditional toxicological tests using conventional methods have thus led to the erroneous conclusion that the low levels of EDCs to which wildlife and humans are exposed do not result in adverse effects. We have worked to establish a new test method for risk assessment of EDCs based on systematic monitoring of receptor signaling and gene expression profiling. In this report, I will present a brief overview of some of our progress.

Using recent developments of DNA microarray technology, the expression and/or suppression of tens of thousands of genes can be measured simultaneously. This will have a big impact on EDC research, with the potential to increase the specificity of individual EDC classification. Gene expression profiles can be used for screening and testing risk assessment. Developed and validated methods for risk assessment of EDCs using DNA microarray technology could provide alternatives to traditional animal bioassays that would be much faster, more sensitive, and more informative.

The EDCs that we have tested were listed in the SPEED '98 report, and were selected for priority risk assessment in 2000 by the Exogenous Endocrine Disrupting Chemicals Task Force. We tested seven chemicals out of eight, which are tributyltin (TBT), 4-octylphenol (OP), nonylphenol (NP), di-n-butylphtalate (DBP), octachlorostyrene (OCS), benzophenone (BP), di-cyclohexylphthalate (DCHP), di-(2-ethylhexyl)phthalate (DEHP), diethylstilbestrol (DES), ethinylestradiol (EE), and 17β-estradiol (E2) called them "test chemicals" here.

We have examined the influence of the test chemicals on the steroid -hormone-sensitive cell lines MCF7, T-47D, TM3, and TM4 using DNA microarrays. MCF7 and T-47D are ER (α & β) positive human breast cancer cell lines. TM3 is an ER (α & β) positive mouse Leydig cell line. TM4, mouse Sertoli cell line, is an ER (β) and an androgen receptor (AR) positive. Concentrations of test chemicals were determined by cell growth assay experiment. Exposure time was 6 hours after adding the test chemicals in the culture medias. Total RNA was isolated by phenol-chloroform extraction or binding/washing method to a silica-based fiber matrix, and mRNA was purified by magnetic separation. Cy3- and Cy5-labeled probes were prepared using purified mRNA from control or exposed cells, respectively (Fig. 1). Labeled probes were hybridized to the Takara IntelliGene microarrays (Human CHIP 1K; 1152 genes, Human DNA CHIP for endocrine disruption study; 432 genes, and Mouse CHIP Set I; 1152 genes) and to the Incyte's microarrays (Human UniGEM V2.0; 8556 genes and Mouse GEM 1; 7634 genes). Generally microarray hybridization and analysis were done according to the procedures of Brown's lab. (http://cmgm.Stanford.edu/pbrown). Fluorescent images of hybridized microarray were obtained by using an Affymetrix 428 array scanner with ImaGene software or an Axon GinePix 4000 scanner with GEMTools software.

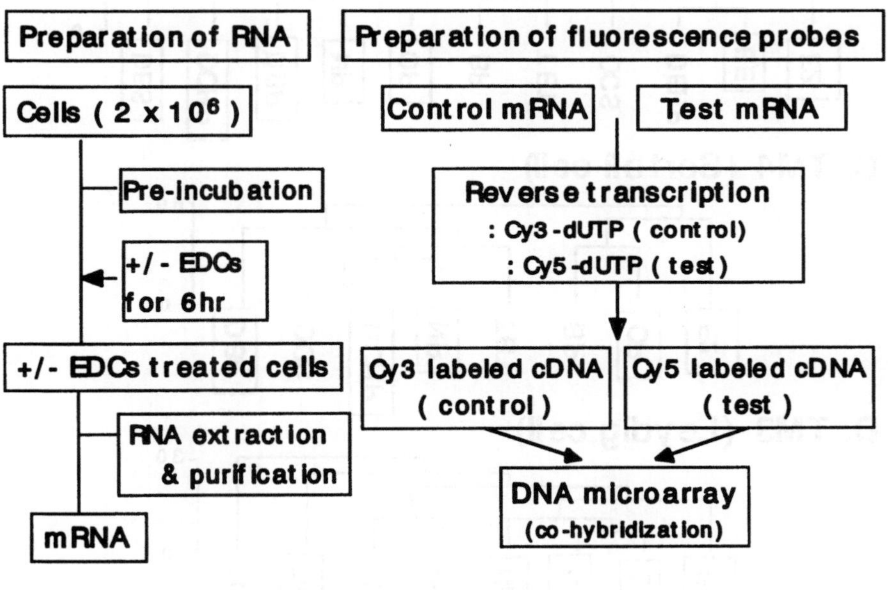

Fig. 1. Preparation of probes for DNA microarray hybridization.

Many genes were influenced significantly at set dose of the test chemicals. There are clear differences in the expression pattern of genes up- or down-regerated by the test chemicals. Gene expression patterns of four cell lines, which exposed to test chemicals, were analyzed by hierarchical clustering using GeneSight software. Results in Fig. 2 were displayed by the clustering dendrograms. NP, OP, DBP, DCHP, DES, EE, and E2, which were confirmed estrogenic activity by E-screen assay, are indicated by box (Fig. 2). Expression of ER α & β are known in MCF7, T47D, and TM3 and the clustering dendrograms showed a similar pattern.

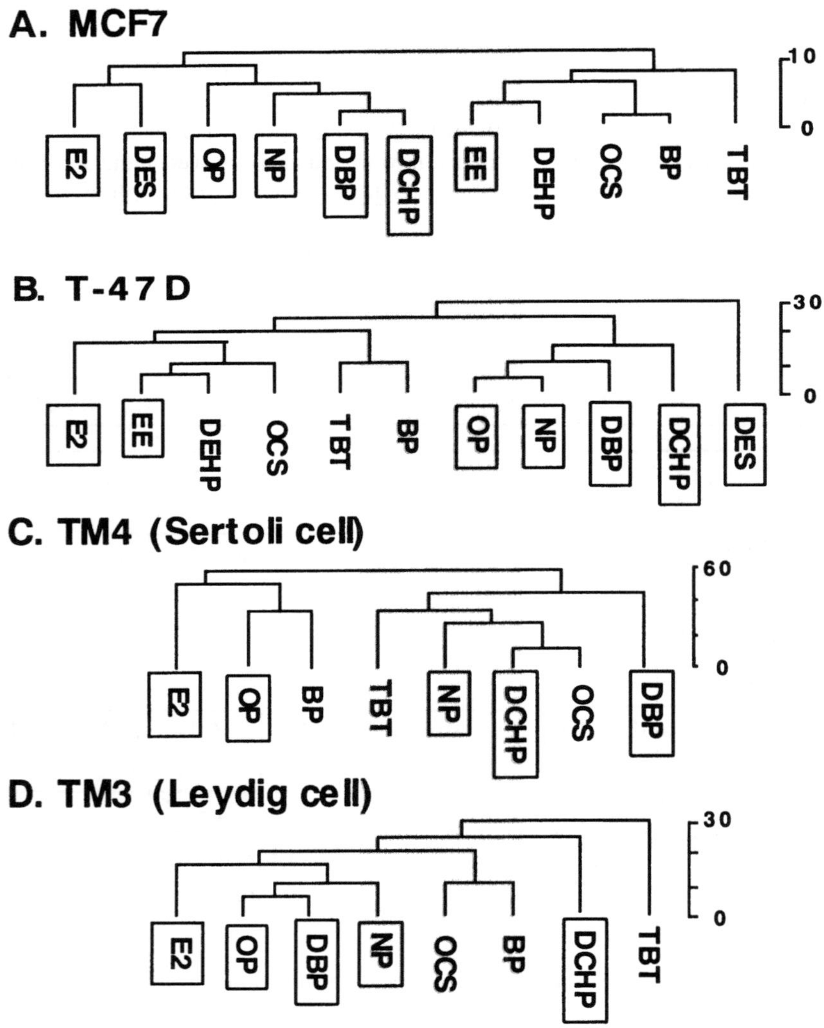

Fig. 2 (A-D). Clustering results of cell lines treated with test chemicals

We identified many genes, which were influenced to chemicals, and we are currently preparing microarrays (EDC chip) including selected expression regulated genes by EDCs, both for humans and mice. In the future, to demonstrate the utility of the EDC chip for risk assessment of EDCs, we will characterize the sensitivity and reproducibility of gene expression profiling tests using pharmacological approaches, e.g. uterotrophic assay. The goal includes establishment of test methods, classification of expression profiles, and construction of databases on individual EDCs using the EDC chip.

Our research was funded in part by a grant from the Ministry of the Environment, Japan.

References

Bolger R, Wiese TE, Ervin K, Nestich S, Checovich W (1998) Rapid screening of environmental chemicals for estrogen receptor binding capacity. Environ Health Perspect 106:551-557

Colborn T, vom Saal FS, Soto AM (1993) Developmental effects of endocrine-disrupting chemicals in wildlife and humans. Environ Health Perspect 101: 378-384

Gordon MN, Osterberg HH, May PC, Finch CE (1986) Effective oral administration of 17 β -estradiol to female C57BL/6J mice through the drinking water. Biol Reproduct 35:1088-1095

Harrison PTC, Holmes P, Humfrey CDN (1997) Reproductive health in humans and wildlife: are adverse trends associated with environmental chemical exposure. Sci Total Environ 205:97-106

Howdeshell KL, Hotchkiss AK, Thayer KA, Vandenberg JG, vom Saal FS (1999) Exposure to bisphenol A advances puberty. Nature 401:763-764

Metzger D, White JH, Chambon P (1988) The human oestrogen receptor functions in yeast. Nature 334:31-36

Nagel SC, vom Saal FS, Thayer KA, Dhar MG, Boechler M, Welshons WV (1997) Relative binding affinity-serum modified access (RBA-SMA) assay predicts the relative in vivo bioactivity of the xenoestrogens bisphenol A and octylphenol. Environ Health Perspect 105:70-76

Nishikawa J, Saito K, Goto J, Dakeyama F, Matsuo M, Nishihara T (1999) New screening methods for chemicals with hormonal activities using interaction of nuclear hormone receptor with coactivator. Toxicol Appl Pharmacol 154:76-83

Reel RJ, Lamb IV JC, Neal BH (1996) Survey and assessment of mammalian estrogen biological assay for hazard characterization. Fundament Appl Toxicol 34:288-305

Soto AM, Sonnenschein C, Chung KL, Fernandez MF, Olea N, Serrano FO (1995) The E-SCREEN assay as a tool to identity estrogens: an update on estrogenic environmental pollutants. Environ Health Perspect 107:113-122 vom Saal FS, Cooke PS, Buchanan DL, Palanza P, Thayer KA, Nagel SC, Parmigiani S, Welshons WV (1998) A physiologically based approach to the study of bisphenol A and other estrogenic chemicals on the size of reproductive organs, daily sperm production, and behavior. Toxicol Ind Health 14:239-260

Application of cDNA microarray for uterotrophic assay

Kwong-Kwok Wong,[1] Jun Kanno,[2] Rita Cheng,[1] Lyle Sasser,[3] James Morris,[3] Larry Anderson,[3] Joel Pounds,[3] and Tohru Inoue[2]

[1]Hematology/Oncology Section, Department of Pediatrics, Texas Children's Cancer Center, M3-3320, Baylor College of Medicine, Houston, Texas 77030, USA
[2]Division of Cellular and Molecular Toxicology, National Institute of Health Sciences, Tokyo, Japan
[3]Battelle Pacific Northwest National Laboratory, Richland, Washington, USA

Summary. To develop a sensitive cDNA microarray based uterotrophic assay, ovarectomized mice were treated with a low dose of 17-β-estradiol (0.2 μg/kg/day) over a three-day period. The average increases in uterine weight were 13%, 23% and 70% after treatment at day 1, day 2 and day 3 respectively. Twenty-four hours after each treatment, uteri were dissected for total RNA extraction and gene expression profiles were assayed with a mouse cDNA microarray containing more than 5000 cDNA elements. From the analysis, we were able to detect 72 genes that were induced more than 2-fold 24 hours after the ovarectomized mice were treated a single dose of 17-β-estradiol. 49 of these genes form a tight cluster when analyzed by the software OmniVizPro™ based on their temporal expression profiles. The number of genes induced more than two-fold increases to more than 200 after the ovarectomized mice were treated with 17-β-estradiol once a day for 1 or 2 more days. These inducible genes include both known and unknown genes. Identified known genes are involved in cell division, transcription activation, stress response, oncogene, and other biochemical activities. These results suggest that gene expression profiles can be used as an alternative endpoint for uterotrophic assay. Further analysis and exploitation of this set of genes will allow us to develop a more sensitive and specific assay for the detection of estrogenic chemical as well as the understanding of the signaling pathway elicited by 17-β-estradiol.

Key words. Uterotrophic assay, endocrine disruptors, 17-β-estradiol, microarray, tyramide signal amplification

Introduction

Endocrine disrupting chemicals have been seen around the world, particularly in the aquatic environment in the past decade (Goldman and Tokuru, 2000). Studies of a variety of natural and man-made chemicals have demonstrated that some of these chemicals can interfere normal sexual differentiation and development in both human and animals. Thus, there is an urgent need to develop a sensitive and rapid method for the detection of endocrine disrupting chemicals in the food and environment.

The rodent uterotrophic assay is the most extensively used assay for testing estrogenic activity in vivo (Thigpen et al. 1987). In the assay, ovarectomized or immature rodents are administrated with an unknown chemical consecutively for 3-4 days. Then, the sizes and weight of the uteri from treated rodents are compared to the vehicle controls. An increase in the wet weight of the uterus is indicative of an estrogenic effect (Reel et al. 1996). However, the uterotrophic response of treated rodent depends on the dose of the chemical used. When a low dose of endocrine disruptor or a weekly active compound is used, the change in size or weight of the uterus may not be significant. Moreover, estrogenic effects can also indirectly mediated and these responses would not be detected by the assay. cDNA microarray (Schena et al. 1995) has been applied to many biological problems and will broaden our understanding of estrogenic effects in term of the expression profile of thousands of genes. Since the gene expression change in the context of the uterotrophic assay is still poorly understood, the aim of this study is to identify 17-β-estradiol inducible genes in ovarectomized mouse at low dose. Hopefully, a more sensitive method based on gene expression profiles can be developed to monitor estrogenic chemicals in the environment or to test the toxicity of man-made chemicals.

Material and Methods

Uterotrophic assay. This study was conducted according to the NIH Guidelines for the Care and Use of Laboratory Animals. The animal protocol was approved by the Institutional Animal Care and Use Committee of Battelle Northwest. Mature 6-weeks old ovarectomized CD1 mice were obtained from Charles River Laboratory, Raleigh, NC for this study. The mice were housed 3/cage in the animal facility for the duration of the studies and allowed access to food and water. All animals were fed with phytoestrogen low diet (PLD) (Table 1) for two weeks before treatment. PLD has no detectable genistein, genistin, daidzein and daidzin (data not shown), which have estrogenic activities. The animal rooms were environmentally controlled for temperature and humidity and maintained on a 12:12 hour light/dark cycle. 17-β-estradiol was prepared as a 2mg/ml stock solution in ethanol and then diluted in sesame oil to 0.04 μg/ml for injection. About 0.15ml diluted 17-β-estradiol solution was used in each injection. Approximately 12 mice in a group of three were injected subcutaneously with 17-β-estradiol at a dose of 0.2 μg/kg once a day for one, two or three consecutive days. Another 12 mice were injected with appropriate vehicle and served as controls. At 1, 2, 3, and 7 days of the experiment (24 hrs after the last dosing), 3 mice from each treatment and control group were killed by CO_2 euthanasia. The uterus was gently removed from the body and the mesometrium and fat were trimmed from the uterus. The uterus was immediately weighed after trimming and blotting. Uteri from 3 or 4 mice were pooled and immediately frozen in liquid nitrogen and stored in RNAlater solution (Ambion Inc., Austin, TX) for storage before RNA extraction.

Table 1. Composition of pytoestrogen low diet (PLD)

Ingredient (%)	NIH-07 (orginal)	PLD (used in this study)
Dried Skim Milk	5.0	-
Fish Meal 60% protein	10	-
Fish Meal 65% protein	-	14
Soybean Meal 49% protein	12.0	-
Alfalfa Meal	4.0	-
Corn Gluten Meal	3.0	-
Ground Shelled Corn	24.5	8.0
Ground Hard Winter Wheat	23	28.5
Wheat Middings	10	-
Wheat (Flour)	-	40.62
Brewers Dried Yeast	2.0	2.0
Dried Molasses	1.5	-
Molasses	-	0.75
Soy Oil	2.5	-
Corn Oil	-	2.5
Salt	0.5	0.33
Dicalcium phosphate	1.25	1.25
Ground Limestone	0.5	-
Mineral and vitamin premixes	0.25	2.05
Total	100	100

All numbers are given as percentages

RNA extraction. RNA was extracted from uterine tissue using the RNeasy Midi kit (Qiagen, Inc., Valencia, CA) according to the manufacturer's specification. Uteri in RNA later solution were thawed and homogenized in RLT buffer supplied with the RNeasy Midi kit. Extracted total RNA was quantitated by spectrophotometer.

Microarrays, Probes, Hybridization, and Scanning. The mouse array is composed of 5184 mouse cDNA elements, 96 elements of 3 *Arabidopsis* ESTs as spiked controls, and 48 negative control. Of these, 2048 clones have annotations and the rest are just ESTs. All clones were obtained from Research Genetics (Huntsville, AL) as sequence-verified clones. PCR products from these clones were prepared and printed onto CMT-GAPS slides (Corning Inc., Acton, MA) using PixSys 5500XL, a robotic arrayer from Cartesian Technologies Inc. (Irvine, CA). Probes were generated using MICROMAX TSA Labeling and Detection Kit (s Elmer Life Sciences, Boston, MA) from 1 μg total RNA according to manufacturer's instruction. Hybridization signals were detected with ScanArray3000 (s Elmer Life Sciences, Boston, MA) dual laser scanner.

Data Analysis. Hybridization signals from each individual microarray were analyzed with software Imagene™ 4.2 (Biodiscovery Inc., Marina del Rey, CA). Cluster analysis and data mining of temporal gene expression profiles of uteri were performed with software OmniVizPro™ v1.6 (OmniViz Inc., Maynard, MA).

Results and Discussion

We have fabricated a mouse cDNA microarray with more than 5000 mouse cDNA for this study. Ovarectomized CD-1 mice were treated with 17-β-estradiol at a dose of 0.2 μg/kg once a day over a three days period. The dose of 17-β-estradiol used was 10 times less than the commonly used dose (Carthew et al. 1999). Uteri were dissected 24 hours after 1, 2 and 3 days of treatement. Additionally, uteri were also dissected from mice at day 7 after the last treatment at day 3. The average wet weight of the uterus increases about 13%, 23%, and 70% after administration of 17-β-estradiol once a day for 1, 2 and 3 days of treatment respectively. The wet weight of the treated mice returned to that of vehicle controls at day 7 after the last treatment at day 3.

Since only minute amount of uterine tissue was available from each mouse, a tyramide signal amplification method (Wong et al. 2001), which only requires 1 μg total RNA, was adopted to generate robust hybridization signals for cDNA microarray analysis. Total RNAs extracted from pooled uteri from treated mice were compared with vehicle controls at the corresponding days. Expression profiles of uterine tissues were generated from mice during different time points in the uterotrophic assay. After normalization, expression data were pre-processed by data filtering on the basis of expression threshold and signal/background ratios. Subsequently data mining and clustering analyses were done with the software OmniVizPro™ (www.omniviz.com).

From the analysis, we were able to detect 76 genes that were induced more than 2-fold after one day of 17-β-estradiol treatment. The number of genes induced more than two-fold increases to more than 200 genes after 2 or 3 days of 17-β-estradiol treatment (Figure 1).

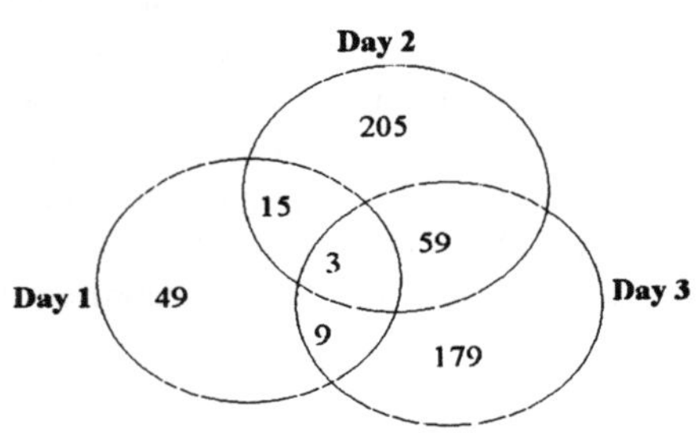

Fig. 1. Summary of temporal changes in the number of 17-β-estradiol inducible genes. Number of inducible genes in different temporal periods are shown.

These inducible genes include both known and unknown genes. Genes involved in cell division, transcription activation, stress responses, oncogene, and other biochemical activities were identified as inducible genes. Clustering analysis of gene expression profiles revealed two major temporal gene expression patterns. The first set of genes were induced more than 2-fold at day 1, but the fold-induction decreased at day 2 and day 3 (Table 2). The pattern of gene expression suggests that a feedback mechanism may control the induction of these early inducible genes. Two chaperonins, calcium binding protein, G protein, and homeobox genes are members of these early-inducible genes. Estrogen has shown to modulate G protein receptors (Kelly and Wagner, 1999). On the other hand, another set of genes has higher induction at day 2 or day 3 (Table 3). Besides those unknown genes, genes related to two major biological functions could be identified among these inducible genes.

The first class of genes may be related to the metabolism of hormones. Since sepiapterin reductase shares high homology with estradiol 17 beta-dehydrogenase

Table 2. List of early inducible known genes with highest induction ratios at day 1.

Accession number	Name	Day 1	Day 2	Day 3
AI327085	heat shock 10 kDa protein 1 (chaperonin 10)	5.7	2.6	1.9
AI326170	NADH dehydrogenase 1 alpha subcomplex 2	3.8	1	1.4
AI326778	homeo box A5	3.4	1.2	1.6
AI326773	histidine rich calcium binding protein	3.3	2.4	2
AI449766	Weakly similar to estrogen-responsive finger protein	3.2	2.7	1.9
AI327155	transforming growth factor beta 1 induced transcript 1	3.1	1.7	1.6
AI324640	S-adenosylmethionine decarboxylase 1	3.1	1.3	0.8
AI323639	solute carrier family 35 (UDP-galactose transporter)	3.1	2.3	1.4
AI323666	cyclin-dependent kinase 2	3	1.7	0.9
AI894315	guanine nucleotide binding protein (G protein)	2.8	1.3	1.2
AI324697	small nuclear ribonucleoprotein polypeptide A	2.7	1.6	1.6
AI327204	transcriptional regulator, SIN3B (yeast)	2.6	1.9	1.5
AI324139	MAD homolog 2 (Drosophila)	2.6	2.4	1.8
AI324947	chaperonin subunit 6a (zeta)	2.6	1.9	1.3
AI326091	peroxiredoxin 4	2.4	2.1	2
AI415665	calumenin	2.4	1.7	2
AI415181	phosphatidylinositol 3 kinase, regulatory subunit	2.3	2	1.4
AI414288	WW domain binding protein 5	2.1	1.9	1.3

The gene induction ratios for each day are the ratios between gene expressions of the 17-β-estradiol treated mice and vehicle control at the corresponding day. Total RNA was pooled from 3 treated mice and vehicle controls for comparison.

Table 3. List of late inducible genes with higher induction ratios at day 2 or day 3.

Accession number	Name	Day 1	Day 2	Day3
AI327152	cell division cycle 20 homolog (S. cerevisiae)	2.2	3.0	6.2
AI894147	sepiapterin reductase	1.7	3.0	5.8
AI428398	antigen identified by monoclonal antibody Ki 67	1.8	2.2	4.3
AI326777	clade H (heat shock protein 47), member 1	2.6	2.4	3.5
AI894225	drebrin 1	1.1	4.1	3.2
AI327173	amyloid beta (A4) precursor-like protein 1	1.5	2.9	3.1
AI413963	laminin receptor 1 (67kD, ribosomal protein SA)	0.6	2.5	3.0
AI426251	tumor-associated calcium signal transducer 2	1.1	2.6	2.9
AI385633	chromogranin B	1.5	2.6	2.9
AI324131	Similar to v-Ki-ras2 viral oncogene homolog	1.3	2.4	2.8
AI450616	claudin 8	0.6	2.1	2.8
AI451127	solute carrier family 3, member 1	1.2	2.9	2.8
AI448863	protein related to DAC and cerberus	0.8	2.6	2.7
AI326265	dynein, axon, heavy chain 11	1.4	3.5	2.5
AI894247	tubulin, beta 5	1.4	2.9	2.5
AI449928	Proteasome 26S subunit, non-ATPase, 8	1.5	2.9	2.3
AI415723	staphylococcal nuclease domain containing 1	1.3	3.3	2.3
AI450786	lymphocyte antigen 94	1.3	4.2	2.3
AI414814	male enhanced antigen 1	1.1	7.3	2.2
AI449363	globin inducing factor, fetal	1.5	4.3	2.2
AI450851	RNA polymerase II transcriptional coactivator	2.4	2.2	2.2
AI415313	calcium channel beta 3 subunit	0.9	2.5	2.1
AI326287	tubulin alpha 8	1.2	2.7	2.0

Only known genes are listed.

(Ichinose et al. 1991), it is tempting to speculate that sepiapterin reductase may be involved in the metabolism of 17-β-estradiol. Besides sepiapterin reductase, the chromogranins are believed to play a role in hormone packaging within secretory granules, in hormone secretion, and serve as prohormones for various proteolytic cleavage products. Furthermore, the mRNA levels of different chromogranins are regulated by estrogen (Lloyd et al. 1992).

The second class of genes is related to tissue remodeling of the cytoskeleton. Heat shock protein 47 (Hsp47) is a procollagen/collagen specific molecular chaperone protein derived from the serpin-family of proteins and essential for the early stages of collagen biosynthesis (Nagata, 1996). Drebrin and tropomyosin competitively bind to actin filaments to regulate the actin remodeling (Ishikawa et al. 1994) Claudins are responsible for tight junction specific obliteration of the intercellular space (Kubota et al. 1999) Beta- and alpha-tubulin, involved in microtubules formation, have also been shown to be regulated by estrogen in ovarectomized rat (Joseph et al. 1982).

The application of cDNA microarray approach to uterotrophic assay has identified many 17-β-estradiol inducible genes. By better understanding of the signaling pathways elicited by 17-β-estradiol, we will be able to detect and classify endocrine disruptors based on what cellular functions that they are mimicking the natural estrogen at low dose. Further analysis of the dose responsiveness of these sets of genes will allow us to develop a more sensitive and specific assay for the detection of estrogenic chemical as well as the understanding of the signaling pathway elicited by 17-β-estradiol.

References

Altucci L, Addeo R, Cicatiello L, Germano D, Pacilio C, Battista T, Cancemi M, Petrizzi VB, Bresciani F, Weisz A (1997) Estrogen induces early and timed activation of cyclin-dependent kinases 4, 5, and 6 and increases cyclin messenger ribonucleic acid expression in rat uterus. Endocrinology 138:978-984

Carthew P, Edwards RE, Nolan BM (1999) Uterotrophic effects of tamoxifen, toremifene, and raloxifene do not predict endometrial cell proliferation in the ovariectomized CD1 mouse. Toxicol Appl Pharmacol 158:24-32

Goldman LR, Koduru S (2000) Chemicals in the environment and developmental toxicity to children: a public health and policy perspective. Environ Health Perspect 108:443-448

Ichinose H, Katoh S, Sueoka T, Titani K, Fujita K, Nagatsu T (1991) Cloning and sequencing of cDNA encoding human sepiapterin reductase--an enzyme involved in tetrahydro-biopterin biosynthesis. Biochem Biophys Res Commun 179:183-189

Joseph MK, Fernstrom MA, Soloff MS (1982) Switching of beta- to alpha-tubulin phosphorylation in uterine smooth muscle of parturient rats. J Biol Chem 257:11728-11733

Kelly MJ, Wagner EJ (1999) Estrogen Modulation of G-protein-coupled Receptors. Trends Endocrinol Metab 10:369-374

Lloyd RV, Hawkins K, Jin L, Kulig E, Fields K (1992) Chromogranin A, chromogranin B and secretogranin II mRNAs in the pituitary and adrenal glands of various mammals. Regulation of chromogranin A, chromogranin B and secretogranin II mRNA levels by estrogen. Lab Invest 67:394-404

Nagata K (1996) Hsp47: a collagen-specific molecular chaperone. Trends Biochem Sci 21:22-26

Reel JR, Lamb IV JC, Neal BH (1996) Survey and assessment of mammalian estrogen biological assays for hazard characterization. Fundam Appl Toxicol 34:288-305

Schena M, Shalon D, Davis RW, Brown PO (1995) Quantitative monitoring of gene expression patterns with a complementary DNA microarray. Science 270:467-470

Thigpen JE, Li LA, Richter CB, Lebetkin EH, Jameson CW (1987) The mouse bioassay for the detection of estrogenic activity in rodent diets: I. A standardized method for conducting the mouse bioassay. Lab Anim Sci 37:596-601

Wong KK, Cheng RS, Mok SC (2001) Identification of differentially expressed genes from ovarian cancer cells by MICROMAX cDNA microarray system. Biotechniques 30:670-675

Large-scale gene expression analysis for evaluation of endocrine disruptors

Hajime Watanabe,[1] Atsuko Suzuki,[1] Takeshi Mizutani,[1] Hiroshi Handa,[2] and Taisen Iguchi[1]

[1]Center for Integrative Bioscience, Okazaki National Research Institutes, CREST and JST, 38 Nishigonaka, Myodaiji, Okazaki 444-8585, Japan
[2]Frontier Collaborative Research Center, Tokyo Institute of Technology 4259 Nagatsuta, Yokohama 226-8503, Japan

Summary. The capacity of numerous chemicals released into the environment to disrupt the development and function of the endocrine system of wildlife and humans is drawing public attention. Early evidence that estrogenic chemicals could pose a threat to human health during development came from studies of diethylstilbestrol (DES) used to prevent premature birth and spontaneous abortion. Laboratory experiments have demonstrated that exposure of animals to sex-hormones during perinatal life cause permanent and irreversible alterations to the endocrine and reproductive systems as well as the immune system, nervous system, bone, muscle and liver in both sexes.
In vitro studies using estrogen receptor expressing cell lines and yeast systems revealed that some chemicals are estrogenic and they can activate transcription of the genes that have estrogenic activity. Based on these results, it has been widely asserted that xenoestrogens affect via estrogen receptor and evoke estrogenic effects in wildlife and humans, but little is known about the disruption mechanisms.
In order to examine the hypothesis, we used DNA microarray to analyze global uterine gene expression patterns in mice. Mice were treated with various chemicals proposed to have endocrine disrupting qualities. Gene expression changes by the chemicals were compared to those by estrogen. As a result, we have found characteristic gene expression patterns for each chemical that are distinct from that of estrogen. We showed that, based on these expression patterns, estrogenicity of chemicals could be estimated. This analysis provides a new method for evaluation of estrogenic chemicals distinct from simple estrogen binding receptor estimation.

Key words. Endocrine disruptors, DNA microarray, Estrogen, Mouse, Uterus

Introduction

Toxicity of chemicals has long been estimated by their lethality, i.e.LD50, until the arising of the endocrine disruption problem. Realization of endocrine disruptors is changing the notion of toxicity because endocrine disruptors are thought to have hazardous effect on the endocrine system with a very low concentration compare to LD50. It indicates that some of chemicals have hazardous effects at the gene expression level without affecting lethality. Thus, a novel approach from the standpoint of gene expression is important for precise evaluation of toxicity. In this meaning, the study of endocrine disruptors can be a good model for toxicogenomics. While we have to deal with various chemicals and target organs in toxicogenomics, the number of suspected chemicals is smaller and target organs are mainly restricted those related to reproduction in the study of endocrine disruptors.

It is widely recognized that many chemicals released into the environment disrupt the endocrine system in wildlife and humans. In laboratory animals, mice exposed perinatally to estrogens provide a model for exploration of the consequences of diethylstilbestrol (DES) exposure in the human, because mouse genital tract development at birth is similar to that of the human fetus at the end of the first trimester. Exposure of animals to sex hormones during perinatal life cause permanent and irreversible alterations to the endocrine and reproductive systems as well as the immune system, nervous system, bone, muscle and liver in both sexes(McLachlan, 2001)

In addition to DES, low levels of xenoestrogens in the environment have disturbed reproduction and the developmental processes in animals. It was realized that a variety of chemicals suspected to have estrogenic activity are present in the environment. They range from synthetic compounds such as nonylphenol (Soto, et al., 1991), bisphenol A (Krishnan, et al., 1993), DDT (Bitman, et al., 1968) to natural phytoestrogens such as Genistein and Daizein.
Based on the notion that some endocrine disruptors are thought to mimic estrogen action, many systems to evaluate estrogenicity of the chemicals have been developed.

The simplest system was competitive binding of chemicals to estrogen receptor using radio labeled estradiol. In addition to the traditional system, conventional systems have also been developed. In a yeast system, the estrogen receptor gene and a reporter gene that has an estrogen responsive element (Coldham, et al., 1997) in its promoter region are introduced into a yeast cell. In this system, estrogen receptor binding activities of chemicals are estimated from the activity of the reporter gene. Another system, called E-SCREEN, is using MCF-7 cells that are expressing estrogen receptor (Soto, et al., 1995). As the growth of the cell is dependent on estrogen concentration, estrogenicity of chemicals can be estimated from growth curve of MCF-7.

These systems are very powerful but only give us binding activities of chemicals to estrogen receptor and it doesn't give any information about actual effects of these chemicals.

As there is no direct evidence indicating that estrogen receptor is responsible for mediating the diverse effects of endocrine disruptors, only the binding information is not sufficient for understanding of the endocrine disruption problem. As it is well known, the estrogen receptor is a transcription factors, analysis of changes in mRNA expression patterns caused by estrogen and

endocrine disruptors can be a powerful tool for understanding of molecular mechanisms underlying endocrine disruption. To analyze genome-wide gene expression changes using uterine tissue, we employed a DNA microarray technique. DNA microarray technology has recently been developed and successfully applied for genome-wide analysis of gene expression as influenced by various stimuli such as serum (Iyer, et al., 1999), hormone (Feng, et al., 2000) or chemicals (Marton, et al., 1998). High-density oligonucleotide arrays (Lockhart, et al., 1996) are especially suitable for genome-wide mapping of gene expression because a large number of genes can be analyzed at one time.
By checking the genes expression pattern affected by estradiol and endocrine disruptors, we can evaluate the chemicals in a gene expression level.

Materials and Methods

Animals

Animals were housed under a 12-h light, 12-h dark cycle. C57/BL6/J mice were ovariectomized at eight weeks of age and treated with 17b-estradiol or chemicals at two weeks postovariectomy. Six hours after the administration, uterus was disected.

Preparation of labeled cRNA and hybridization

Total RNA was extracted from uterus by using TRIZOL (LifeTechnologies) and purified by using RNeasy (Qiagen). Total RNA was converted into double stranded cDNA by using SuperScript Choice System (LifeTechnologies) with an oligo-dT primer contaning a T7 RNA polymerase promoter (APB). Using purified double-stranded cDNA, in vitro transcription was performed by using a T7 Megascript Kit (Ambion) in the presence of biotin-labeled nucleotides (Enzo Diagnostics). Generally, 20 μ g of total RNA was used for cDNA synthesis and a half of the product was used to make biotin-labeled cRNA. Labeled cRNA was purified by using RNeasy (Qiagen) and fragmented to sizes ranging from 35 to 200bases by incubating at 94°C for 35 min.

Fragmented cRNA was mixed with hybridization buffer containing 100mM MES, 1M NaCl, 20mM EDTA and 0.01% Tween 20 and control oligonucleotides. High-density oligonucleotide array (Mouse U74 A) was obtained from Affymetrix (APB) and hybridization was performed at 45°C for 16hr. After hybridization, arrays were washed, stained with streptavidin-phycoerythrein (Molecular Probes) by using Affymetrix fluidix station and scanned with argon-ion laser confocal scanner (APB).

Data analysis

Scanned data were analyzed with the GeneChip software (Affymetrix) and detailed protocols for data analysis have been described (Lockhart, et al., 1996). Briefly, each gene is represented by the use of 20 perfectly matched (PM) and mismatched (MM) control oligonucleotides. The MM probes are used to detect background level and cross-hybridization signals. To determine the quantitative RNA abundance, the average of the differences representing PM minus MM for each gene specific probe is calculated. For normalization of average differences, rescaling factors were calculated to give average intensity as 2500 in each experiment.

Results and Discussion

Evaluation of chemicals from gene expression pattern

In order to evaluate effects of chemicals on gene expression, we compared fold change of each gene expression level between estradiol administration and chemical administration.

In this study, we used several chemicals including phytoestrogen, Diethylstilbestrol, nonylphenol, bisphenolA, Daizein and Genistein.

The fold change of each gene expression was calculated by the ratio of average difference between chemical treated and oil treated samples and this value was used to compare gene induction pattern between chemical treated and estradiol treated samples by scatter plot (Fig. 1).

Suppose that gene activation patterns were exactly equal between estradiol and chemical, each gene would have the same fold change value and would be spotted on a diagonal line. Thus the slope of the regression line indicates the similarity of gene expression pattern. In addition, R square values derived from the regression line give information of similarity of gene expression pattern. These values are also indicated in Fig.1. It was shown that regression line and R square values represents well the estrogenicity of the chemicals because estimation by this method consists well with the uterotrophic estimation (data not shown), indicating that, though this is a simple method, this system could evaluate chemicals.

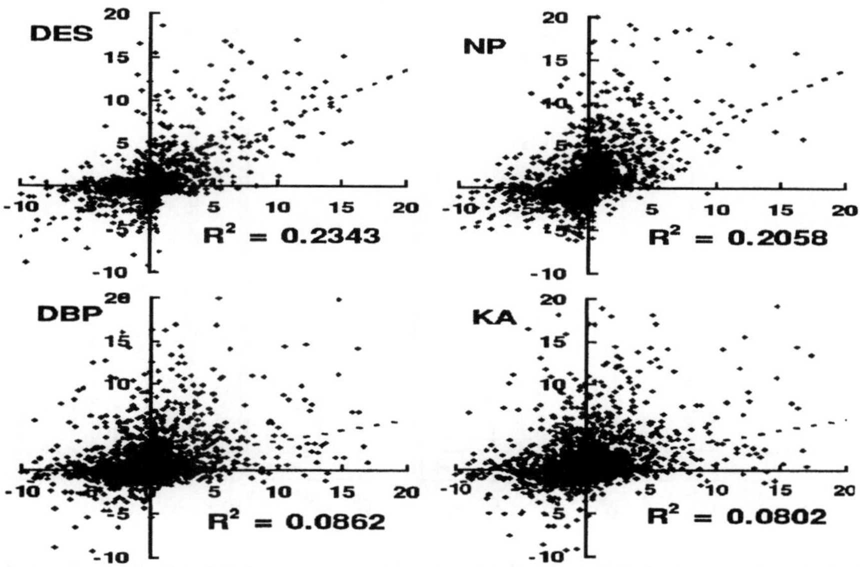

Fig. 1. Scatter plot of fold changes of gene expression level altered by the chemicals and estradiol. X-axis indicated fold changes of gene expression altered by estradiol administration and Y-axis indicated fold changes of gene expression altered by a chemical administration. Dotted line indicates a diagonal line. Equation of regression line and R square value are indicated in each panel.

Evaluation of chemicals using selected genes

To improve the estimation of estrogenicity, we selected genes that were suitable for the estimation. For this purpose, expression analysis of estrogen-induced genes were performed three times and about 2000 genes whose scores were higher by the estimation of Student's t-test were selected. In several genes, expression levels were examined using quantitative RT-PCR and confirmed that they were actually induced by estradiol (data not shown).

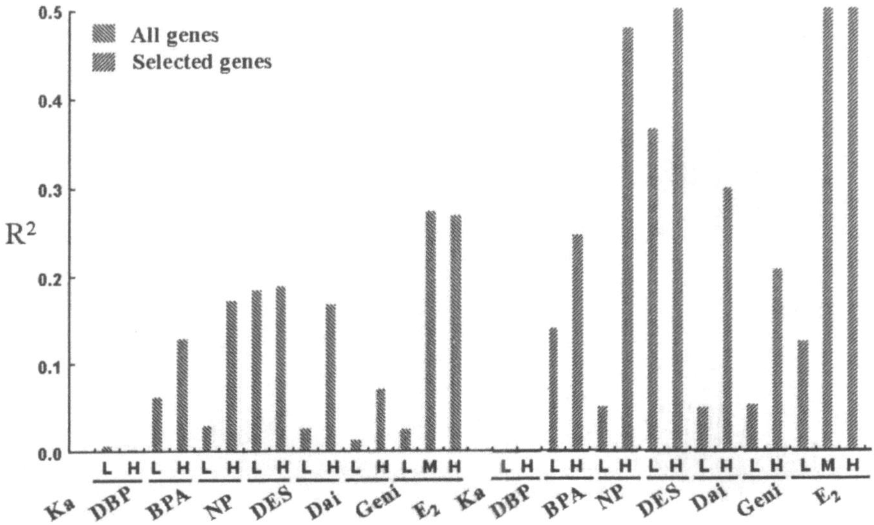

Fig.2. Evaluation of chemicals by R square values.Based on regression lines, R square values were calculated. Left panel shows the values calculated from all genes. Right panel shows the values calculated from the selected genes. Ka, Kanamycin; DBP, dibuthylphthalate; BPA, bisphenol-A; NP, nonylphenol; DES, diethylstilbestrol; Dai, Daidzein; Geni, Genistein; E2, 17b-estradial. L, H, M indicate administrated doses. L, 0.1mg/20g B.W.; H, 1mg/20g B.W. except for DES, Genistein and E2. For DES and Genistein, H indicates 0.1mg and 0.1μg, respectively and L indicates ten times smaller amount. For E2, H, 0.1μg/20g B.W.; M, 0.01μg/20g B.W.; L, 0.001μg/20g B.W.

Using these selected genes, regression lines and R square values are calculated (Fig.2.). We showed that, by selecting genes, the values could be improved because difference of chemicals became clearer. The chemicals that had been reported to have estrogen receptor binding activities, such as diethylstilbestrol and nonylphenol were scored higher. On the other hand, dibuthylphthalate and kanamycin that were thought to be non-estrogenic chemicals had very low R square values. Although it is difficult to make a clear threshold, evaluation of chemicals becomes much easier using selected genes.

To evaluate relationship between these chemicals, we also found that clustering analysis (Eisen, et al., 1998) on selected genes could be used for the evaluation of the chemicals. Information about mutual similarity of chemicals could be obtained from this analysis (data not shown).

In this study, we used two doses of chemicals, and found that the high dose of chemicals except for dibuthylphtalate and kanamycin, have an estrogenic effect. Although precise estimation of chemicals on estrogenicity should be determined in combination with other system, our study showed that evaluation of chemicals could be possible by gene expression analysis.

References

Bitman J, Cecil HC, Harris SJ, Fries GF (1968) Estrogenic activity of o,p'-DDT in the mammalian uterus and avian oviduct. Science 162:371-372

Coldham NG, Dave M, Sivapathasundaram S, McDonnell DP, Connor C and Sauer MJ (1997) Evaluation of a recombinant yeast cell estrogen screening assay. Environ Health Perspect 105:734-742

Eisen MB, Spellman PT, Brown PO, Botstein D (1998) Cluster analysis and display of genome-wide expression patterns. Proc Natl Acad Sci U S A 95:14863-14868

Feng X, Jiang Y, Meltzer P, Yen PM (2000) Thyroid hormone regulation of hepatic genes in vivo detected by complementary DNA microarray. Mol Endocrinol 14:947-955

Iyer VR, Eisen MB, Ross DT, Schuler G, Moore T, Lee JC, Trent JM, Staudt LM, Hudson J, Jr., Boguski MS, Lashkari D, Shalon D, Botstein D, Brown PO (1999) The transcriptional program in the response of human fibroblasts to serum. Science 283:83-87

Krishnan AV, Stathis P, Permuth SF, Tokes L, Feldman D (1993) Bisphenol-A: an estrogenic substance is released from polycarbonate flasks during autoclaving. Endocrinology 132:2279-2286

Lockhart DJ, Dong H, Byrne MC, Follettie MT, Gallo MV, Chee MS, Mittmann M, Wang C, Kobayashi M, Horton H and Brown EL (1996) Expression monitoring by hybridization to high-density oligonucleotide arrays. Nat Biotechnol 14:1675-1680

Marton MJ, DeRisi JL, Bennett HA, Iyer VR, Meyer MR, Roberts CJ, Stoughton R, Burchard J, Slade D, Dai H, Bassett DE, Jr., Hartwell LH, Brown PO, Friend SH (1998) Drug target validation and identification of secondary drug target effects using DNA microarrays. Nat Med 4:1293-1301

McLachlan JA (2001) Environmental signaling: what embryos and evolution teach us about endocrine disrupting chemicals. Endocrine Reviews 22:319-341

Soto AM, Justicia H, Wray JW, Sonnenschein C (1991) p-Nonyl-phenol: an estrogenic xenobiotic released from "modified" polystyrene. Environ Health Perspect 92:167-173

Soto AM, Sonnenschein C, Chung KL, Fernandez MF, Olea N, Serrano FO (1995) The E-SCREEN assay as a tool to identify estrogens: an update on estrogenic environmental pollutants. Environ Health Perspect 103 Suppl 7:113-122.

Analysis of toxicogenomic response to endocrine disruptors in the mouse testis

Masatoshi Komiyama,[1] Tetsuya Adachi,[1,2] and Chisato Mori[1,2]

[1]Department of Bioenvironmental Medicine, Graduate School of Medicine, Chiba University, Chiba 260-8670, Japan
[2]Core Research for Evolutional Science and Technology (CREST), Japan Science and Technology Corporation, Kawaguchi 332-0012, Japan

Summary. In order to evaluate the feasibility of cDNA microarrays for the risk assessment of endocrine disruptors (EDs), alteration of gene expression profiles was analyzed in adult mouse testes after neonatal exposure to diethylstilbestrol (DES), bisphenol A (BPA) or genistein (Gen), using cDNA microarrays. Analysis with Mouse GEM I revealed that the expression levels of 34, 38 and 12 genes were changed in testes of DES-, Gen- and BPA-treated 12-week old mice, respectively. Gene expression profiles were very similar between DES- and Gen-treated mice, but that of the BPA-treated mice was quite different. These results suggest that gene expression profiles might be feasible for the grouping of EDs in terms of their effects. Further, it is suggested that mechanisms of BPA action may be different from those of DES and Gen, although they are all estrogen-like compounds. Next, we investigated the transition of the expression profiles in testes of DES-treated mice using in-house cDNA microarrays. Many of the genes affected by DES were up- or down-regulated at restricted periods, but some genes were continuously up-regulated or shifted from the down-regulated state to the up-regulated state. These results suggest that the transition patterns may represent the gene expression cascades that were affected by EDs. We propose that the cDNA microarray is a useful tool, which provides us a bird's eye view of the global effects of EDs on gene expression.

Key words. DNA microarray, Diethylstilbestrol, Bisphenol A, Genistein, Gene expression profile

Introduction

It has been recently clarified that there are many compounds that disturb functions of natural hormones in the body and they are referred to as endocrine disruptors (EDs). Many of them have the capacity to mimic effects of estrogens and show some adverse effects on the development and/or function of genital organs, particularly when exposure occurs in fetal or neonatal individuals (Mori 2001). Diethylstilbestrol (DES), a synthetic estrogen (Herbst et al. 1974; Arai et al. 1983; Newbold et al. 1984, 1987; Visser et al. 1998), and bisphenol A (BPA), a material of resin-based and plastic products (Brotons et al. 1995), are well known chemical compounds listed as EDs or candidates for EDs. In addition, phytoestrogens such as genistein (Gen) and daizein which are produced by plants, are also known to have estrogenic activity (Kuiper et al. 1998).

Testis is one of the organs that are vulnerable to EDs. Toxicity of chemicals against the testis has been assessed by conventional reproductive and developmental tests using the following parameters; weight and histology of testis, sperm counts and motility, fertility, etc. (Goyal et al. 2001). However, in the case of EDs, these assays are hard and costly because of the time lag between exposure and appearance of detectable effects. Therefore, the establishment of new methods for more sensitive and predictive evaluations of EDs is a growing concern.

DNA microarrays are a powerful and high-throughput tool for monitoring the expression of thousands of genes simultaneously (Schena et al. 1995; Shalon et al. 1996; Duggan et al. 1999; Adachi et al. 2001). If gene expression profiles can be used as a predictive parameter, microarrays will be an ideal tool for toxicological test of EDs. In this study, we analyzed toxicogenomic response to EDs in mouse testes using Mouse GEM I (8800 mouse genes) and in-house cDNA microarrays (2304 mouse genes) to evaluate its feasibility for the risk assessment of EDs.

Materials and methods

Animals

All animals were maintained in accordance with the Chiba University Guidelines for animal experimentation. Timed pregnant female ICR mice were purchased from Japan SLC, Inc. (Hamamatsu, Japan). Mice were allowed access to laboratory diet CE-2 (CLEA Japan, Inc., Tokyo, Japan) and tap water *ad libtum*. Newborn male mice were given daily subcutaneous injections of DES (50 μg/mouse/day; Sigma-Aldrich, Tokyo, Japan), Gen (1 mg/mouse/day; Wako, Osaka, Japan) or BPA (200 μg/mouse/day; Sigma-Aldrich) from neonatal day 1 to day 5 (Iguchi et al. 1986). Sesame oil (Sigma-Aldrich) and olive oil (Sigma-Aldrich) were used as solvents of DES and Gen, respectively. BPA was dissolved in 99.9% ethanol and then emulsified into a mixture of BPA solution, water and olive oil (1:1:3). As controls, only solvents were given. Mice of the Gen-treated group and its control group were fed Gen-free diet consisted of 45.5% cornstarch, 24.5% casein, 10% granulated sugar, 6% corn oil, 5% cellulose powder, 1% α starch, 1% vitamin mixture and 7% mineral mixture (wt/wt). Mice were sacrificed at 4, 8 and 12 weeks of age. Testes were removed and frozen in liquid nitrogen for cDNA microarray analysis.

cDNA microarrays

Total RNA was purified from each frozen tissue using the TRIZOL reagent (Gibco-BRL, Gaithersburg, MD, USA) (Shibayama et al. 2001), and those of 3 testes from the same group were mixed. Poly(A)$^+$ RNA was purified from the total RNA using Oligotex-dT30 (Takara Shuzo Co., Ltd., Kyoto, Japan). For analysis of gene expression, Mouse GEM I (Incyte Genomics, Inc., Palo Alto, CA, USA) and in-house cDNA microarrays (Helix Institute, Kisarazu, Japan) were used. Mouse GEM I is a cDNA microarray, which contains about 8800 cDNA probes. The in-house microarray, containing 2304 cDNA probes prepared from mouse fetuses (day 14.5), was constructed as previously described (Yoshikawa et al. 2000). To generate fluorescent targets, Cy5- or Cy3-dUTP (Amersham-Pharmacia Biotech Japan, Tokyo, Japan) was incorporated during reverse transcription of poly(A)$^+$ RNA, primed by an oligo(dT) primer (Yoshikawa et al. 2000). Mixture of the fluorescent targets was applied onto the microarray surface, and the microarray was incubated overnight in a hybridization chamber at 65°C under a humidified condition. The microarray was washed and then scanned with the use of a fluorescence laser-scanning device (ScanArray4000; GSI Lumonics, Tokyo, Japan).

Results and discussion

Long-term effects of EDs on gene expression in the testis

Using Mouse GEM I, alteration of gene expression profiles was analyzed in testes of 12-week old mice that were neonatally exposed to EDs. The majority of genes examined showed only a small difference in expression and their expression ratios (ED-treated/control) were within a range between 0.5 and 2. However, 2, 3 and 9 genes were up-regulated more than 2-fold, and 32, 35 and 3 genes were down-regulated less than 0.5-fold by neonatal exposure to DES, Gen and BPA, respectively (Adachi et al. 2001).

A clustering analysis of all genes on the microarray using Gene Maths (Infocom, Tokyo, Japan) revealed that gene expression profiles were very similar between DES- and Gen-treated mice, but the expression profile in BPA-treated mice was quite different from the others (Fig. 1). These data suggest that the gene expression profiles might be feasible for the grouping of EDs from a viewpoint of their effects on organs, as has been suggested previously (Bartosiewicz et al. 2001). Further, it is suggested that mechanisms of BPA action may be different from those of DES and Gen, although they are all estrogen-like compounds.

Fig. 1. Clustering analysis of gene expression profiles. Each vertical line represents a single gene expression event. The tree diagrams on the top and right indicate the related genes and data sets, respectively. Gene expression profiles in the testis are similar between DES-treated and Gen-treated mice, but that of BPA-treated mice is quite different from others.

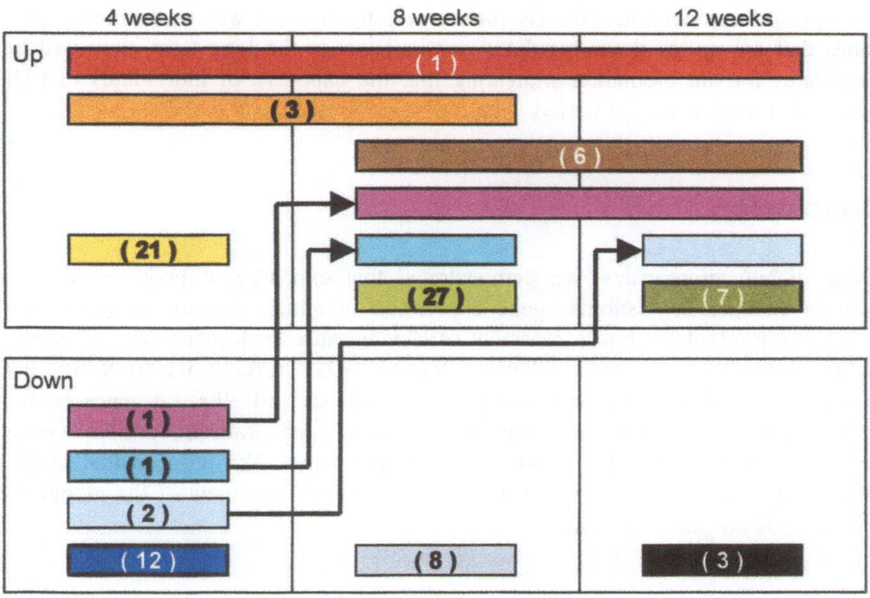

Fig. 2. Transition of gene expression profiles in testes of DES-treated mice during 4–12 weeks. Each rectangle represents the pattern of gene expressions which were altered by neonatal exposure to DES. Some of the rectangles are continuation of others (indicated by *arrows*). Each parenthesized number represents the number of genes contained in the pattern.

Transition of gene expression profiles in testes of DES-treated mice were investigated during 4–12 weeks of age using the in-house microarrays. We found 12 transition patterns of altered gene expressions that were up-regulated more than 1.5-fold or down-regulated less than 0.67-fold (Fig. 2, colored rectangles). There were 25 genes up-regulated at 4 weeks, one of which was highly expressed at 8 and 12 weeks (red), and three of them continued the high expression levels until 8 weeks (orange), while the expression of the others became a similar level as those of the control during 8–12 weeks (yellow). On the other hand, 16 genes were down-regulated at 4 weeks. Out of these genes, 4 genes became highly expressed later; one gene both at 8 and 12 weeks (pink), one gene only at 8 weeks (sky blue), and two genes only at 12 weeks (violet). The other 12 genes were down-regulated only at 4 weeks (blue). There were 27 genes up-regulated (light green) and 8 genes down-regulated (gray) only at 8 weeks, and 7 genes up-regulated (dark green) and 3 genes down-regulated (black) only at 12 weeks.

These results indicate that the gene expression profile gradually changes after neonatal exposure to EDs and the transition patterns may represent the gene expression cascades that were affected by EDs. In this analysis, we found one gene was up-regulated continuously. By further investigations, if we can find more such genes that are up- or down-regulated continuously in the long term, they will be candidates for the biomarker indicating that the exposure of individuals to EDs occurred at some sensitive period.

Conclusion

Using cDNA microarrays, we demonstrated that exposure of neonates to EDs induces changes of testicular gene expression in adults. Identification of such genes is important for risk assessment of EDs because such genes may be useful as biomarkers that can predict adverse effects of EDs. Currently we are testing our new in-house cDNA microarray carrying 800~ mouse and ~900 rat genes related to apoptosis, DNA repair, metabolizing enzymes, signal transduction, etc. Genes responsive to EDs are also included in the microarray. We propose that cDNA microarray is a useful tool, which provides us a bird's eye view of the global effects of EDs on gene expression.

Acknowledgments

This work was done in collaboration with Dr. N. Seki (Helix Research Institute) and Dr. H. Tainaka (Asahi Techno Glass Corporation), and supported by Grants from the following: The Ministry of Education, Science, Sports, Culture and Technology of Japan, New Energy and Industrial Technology Development Organization, Uehara Memorial Foundation, KANZAWA Medical Research Foundation, and Showa Shell Sekiyu Foundation for Promotion of Environmental Research.

References

Adachi T, Sakurai K, Fukata H, Komiyama M, Shibayama T, Iguchi T, Mori C (2001) A DNA microarray analysis for the effect of spermatogenesis to phytoestrogen and endocrine disruptors in mice. Chiba Med J 77:151–158

Arai Y, Mori T, Suzuki Y, Bern HA (1983) Long-term effects of perinatal exposure to sex steroids and diethylstilbestrol on the reproductive system of male mammals. Int Rev Cytosol 84:235–268

Bartosiewicz M, Penn S, Buckpitt A (2001) Applications of gene arrays in environmental toxicology: fingerprints of gene regulation associated with cadmium chloride, benzo(a)pyrene, and trichloroethylene. Environ Health Perspect 109:71–74

Brotons JA, Olea-Serrano MF, Villalobos M, Pedraza V, Olea N (1995) Xenoestrogens released from lacquer coatings in food cans. Environ Health Perspect 103:608–612

Duggan DJ, Bittner M, Chen Y, Meltzer P, Trent JM (1999) Expression profiling using cDNA microarrays. Nat Genet 21:10–14

Goyal HO, Braden TD, Mansour M, Williams CS, Kamaleldin A, Srivastava KK (2001) Diethylstilbestrol-treated adults rats with altered epididymal sperm numbers and sperm motility parameters, but without alterations in sperm production and sperm morphology. Biol Reprod 64:927–934

Herbst AL, Robboy SJ, Scully RE, Poskanzer DC (1974) Clear-cell adenocarcinoma of the vagina and cervix in girls: analysis of 170 registry cases. Am J Obstet Gynecol 119:713–724

Iguchi T, Takase M, Takasugi N (1986) Development of vaginal adenosis-like lesions and uterine epithelial stratification in mice exposed perinatally to diethylstilbestrol. Exp Biol Med 181:59–65

Kuiper GG, Lemmen JG, Carlsson B, Corton JC, Safe SH, van der Saag PT, van der Burg B, Gustafsson JA (1998) Interaction of estrogenic chemicals and phytoestrogens with estrogen receptor beta. Endocrinol 139:4252–4263

Mori C (2001) Possible effects of endocrine disruptors on male reproductive function. Acta Anat Nippon 76:361–368

Newbold RR, Suzuki Y, McLachlan JA (1984) Müllerian duct maintenance in heterotypic organ culture after in vivo exposure to diethylstilbestrol. Endocrinol 115:1863–1868

Newbold RR, Bullock BC, McLachlan JA (1987) Müllerian remnants of male mice exposed prenatally to diethylstilbestrol. Teratogen Carcinogen Mutagen 7:337–389

Schena M, Shalon D, Davis RW, Brown PO (1995) Quantitative monitoring of gene expression patterns with a complementary DNA microarray. Science 270:467–470

Shalon D, Smith SJ, Brown PO (1996) A DNA microarray system for analyzing complex DNA samples using two-color fluorescent probe hybridization. Genome Res 6:639–645

Shibayama T, Fukata H, Sakurai K, Adachi T, Komiyama M, Iguchi T, Mori C (2001) Neonatal exposure to genistein reduces expression of estrogen receptor alpha and androgen receptor in testes of adult mice. Endocrine J (in press)

Visser JA, McLuskey A, Verhoef-Post M, Kramer P, Grootegoed JA, Themmen AP (1998)

Effect of prenatal exposure to diethylstilbestrol on Müllerian duct development in fetal male mice. Endocrinol 139:4244–4251
Yoshikawa T, Nagasugi Y, Azuma T, Kato M, Sugano S, Hashimoto K, Masuho Y, Muramatsu M, Seki N (2000) Isolation of novel mouse genes differentially expressed in brain using cDNA microarray. Biochem Biophys Res Comm 275:532–537

Gene expression profiling to identify mechanisms of male reproductive toxicity

David J. Dix

National Health and Environmental Effects Research Laboratory, Office of Research and Development, U.S. Environmental Protection Agency, Research Triangle Park, NC, 27711, USA.

Summary. With the advent of sequence information for entire mammalian genomes, it is now possible to analyze gene expression and gene polymorphisms on a genomic scale. We have used cDNA microarrays to define the testicular transcriptomes of mice and humans and develop microarrays appropriate for monitoring testicular gene expression. Profiling gene expression in the testis of fertile, subfertile, and infertile mice and humans has been used to highlight genes critical to fertility and responsive to toxicant exposures. Gene expression profiling offers the ability to identify pre-clinical effects of exposure and provide mechanistic data useful for risk assessments, and could ultimately provide a vehicle for developing early diagnostic and preventative measures in at-risk populations or individuals. In animal models, we have used DNA microarrays to identify toxicant mode-of-action as indicated by effects on gene expression relevant to testis histopathology and fertility. After a series of experiments utilizing commercial human and mouse arrays defined the testis transcriptome and identified gene expression changes linked to fertility, we selected 950 mouse genes and created a custom cDNA microarray focused on testicular gene expression. This array has been used to characterize testicular gene expression following exposure of mice to the water disinfectant by-product bromochloroacetic acid in a regimen that induces subfertility at the higher doses. This type of toxicogenomic data has already provided mechanistic data useful for identifying toxicant mode-of-action. In future studies we plan to identify common changes in gene expression between chemicals, tissues, strains and species in order to characterize cumulative and aggregate risks and assist harmonized non-cancer and cancer risk assessments.

Key words. Gene expression, polymorphism, cDNA microarray, reproductive toxicity, risk assessment.

The application of genomics to toxicology

The genomics revolution has been fueled by the explosion of DNA sequence information for numerous organisms. In the post-genomic era, rapidly advancing technologies are available to use this genomic information and examine the sequence, expression, and function of thousands of genes and proteins simultaneously. These technologies represent a paradigm shift from single-gene investigative approaches and are fundamentally altering the practice of every biological science, including toxicology.

Our aim is to improve the scientific foundation of human health risk assessment through the development of methods for characterizing susceptible subpopulations, and through examining the potential utility of emerging technologies in harmonizing risk assessment and determining health status. Gene expression profiling may hold the key to early diagnosis of adverse environmental exposures, as changes in gene expression will always precede clinical manifestation of toxic effects.

Environmental stressors have the potential to affect multiple organizational levels of a cell or tissue:

1. Genome- the chromosomal DNA information.
2. Transcriptome- the mRNA from all actively transcribed genes in a biological sample.
3. Proteome- the entire protein complement of a biological sample.

New genomic and proteomic technologies are being developed to monitor these effects in their totality, enhancing our understanding of the biological sequelae linking exposure to health effects. Thus DNA microarrays can now monitor the total transcriptome, or gene expression of mammalian cells or tissues; and complementary high-throughput proteomics techniques are being developed with the capacity to define the proteome. This global approach to environmental research:

1. encompasses all molecular functions, biological processes and cellular components;
2. bridges the full range of biological effects;
3. requires intensive bioinformatics.

Genomic technologies are sufficiently developed to allow for immediate applications to toxicology (Rockett and Dix, 1999 and 2000). Initiating the use of genomics technologies will lead to dramatic refinement of our understanding of mode-of-action, species/strain differences, and inter-individual variation (i.e., susceptibility) in response to toxicants.

Genomics represents the next step beyond current mode-of-action research, providing data on the complete cellular response to environmental stress. From genomic results, one can distill the key genetic regulators and biological precursors of adverse health effects. After identifying these genetic regulators, additional genomic analyses will better define individual risk through characterization of relevant single nucleotide polymorphisms (SNPs).

Experimental results in reproductive toxicogenomics

Effects of hyperthermia on spermatogenesis, apoptosis, gene expression and male fertility

Gene expression profiling was used to characterize cellular and molecular mechanisms critical for male fertility and responsive to environmental stress (Rockett et al., 2001a). This work expanded on earlier studies of heat shock proteins (HSPs) role in spermatogenesis and protecting testicular cells from environmental hazards such as heat, radiation and chemicals. Cellular and molecular methods were used to characterize effects of testicular heat shock at different times post-treatment. Mating studies confirmed histopathological observations that spermatocytes are the most susceptible cell type. Apoptosis in spermatocytes was confirmed by TUNEL, and was temporally correlated with the expression of stress-inducible Hsp70-1 and Hsp70-3 protein in spermatocytes. To further characterize gene expression networks associated with heat shock induced effects, we used DNA microarrays to interrogate the expression of 2,208 genes and thousands more ESTs expressed in mouse testis. Of these genes, 27 were up-regulated and 151 were down-regulated after heat shock. Array data were concordant with the disruption of meiotic spermatogenesis, the heat-induced expression of HSPs, and an increase in apoptotic spermatocytes. Furthermore, array data indicated increased expression of four additional non-HSP stress response genes, and eight cell adhesion, signaling, and signal transduction genes. Decreased expression was recorded for 10 DNA repair and recombination genes; nine protein synthesis, folding and targeting genes; nine cell cycle genes; five apoptosis genes; and four glutathione metabolism genes. Thus, the array data identify numerous candidate genes for further analysis in the heat shocked testis model, and suggest multiple possible mechanisms for heat shock-induced infertility.

Genomic analysis of the testicular toxicity of haloacetic acids

Gene expression profiles from testes of mice exposed to effective doses of reproductive toxicants can be used to identify genes and gene networks critical to male fertility, infer toxicant mode of action, and understand gene-environment interactions. A custom mouse cDNA microarray interrogating 950 genes (Rockett et al. 2001b) was used to analyze gene expression in the testes of mice exposed to a haloactetic acid (HA). HAs are unintended byproducts of drinking water disinfection. Bromochloroacetic acid (BCA) is a commonly occurring HA that acts as a reproductive toxicant in male rodents, targeting sperm and spermatids. Adult C57BL/6N male mice were dosed via gavage with 0, 8, 24, 72 or 216 mg/kg of BCA once daily for 14 days. The mice were then used in a 40-day sequential breeding assay to determine if BCA negatively affects reproductive performance by targeting a particular phase of spermatogenesis. Decreased fecundity and infertility was observed during the first 10 days of the breeding assay, indicating that

BCA was affecting post-meiotic spermatogenesis. Histological analysis of testes harvested at the end of dosing revealed spermatid retention and abnormal residual bodies, consistent with disruption of spermiation (the process of spermatid separation from Sertoli cells). DNA array analysis of these testes indicated that expression of 42 of the 950 genes interrogated was significantly altered by the higher doses of BCA. Of the 42 altered genes, 10 were cell adhesion genes, 7 stress response genes, and 4 steroid metabolism genes. These 10 cell adhesion-signaling genes included vinculin, a catenin, an integrin and tubulin. These families of proteins are components of or associated with the ectoplasmic specializations (ES) which form cell-cell junctions between Sertoli cells and between Sertolis and spermatids. Thus the gene expression results suggest that BCA exposure disrupts normal ES function and signaling cascades regulating sperm release. Further studies to examine the dose-responsiveness of these effects on gene expression are underway.

Using sperm mRNA as a predictor of fertility and indicator of toxicant effect

We are exploring the feasibility of using gene expression changes in easily accessible or surrogate tissues to see if they mimic gene expression changes taking place in target reproductive tissues. The use of semen as a surrogate tissue to characterize infertility problems in men is being examined, and we have begun characterization and comparison of the transcriptomes of normal adult human testis and sperm. The causes of the reported rise of infertility, testicular cancer and cryptorchidism along with the concomitant decreased sperm count in the ejaculate of individuals from developed economies remains uncertain. To date, a very small subset of genes has been shown to be vital for the development of a healthy, fertile male gamete. Application of microarray technology to the study of male reproductive biology presents a unique opportunity to globally address the mechanisms that control the differentiation of the male gamete during normal, perturbed and diseased states. In collaboration with Drs. Stephen Krawetz (Wayne State University) and David Miller (University of Leeds) we have begun a genome-wide scan to identify and characterize sperm RNA as a marker of male factor infertility and environmental exposure. Using cDNAs derived from RNA isolated from testes and ejaculate spermatozoa, we hybridized a suite of nylon filter cDNA microarrys containing 30,892 Expressed Sequence Tag (EST) probes, of which 27,016 were unique. As expected, all mRNAs isolated from human sperm were present in testis, but all testis mRNAs were not present in sperm. Testes cDNAs hybridized to 7,157 EST probes. In contrast, the fertile spermatozoa cDNAs identified 3,281 ESTs. As expected, the spermatozoal cDNAs identified a discrete subset of the ESTs hybridized by testes cDNAs. If sperm mRNA is predictive of fertility, our next step will be to investigate whether sperm mRNA is also a predictor of adverse effects or indicator of exposure to environmental reprotoxicants.

The use of peripheral blood to monitor adverse effects in the testis

Circumstantial evidence, such as reduction in semen quality and an increase in testicular and prostate cancer, has contributed to speculation that environmental exposures may be linked to a number of adverse reproductive health effects in humans. The challenge exists, therefore, to create a stronger and more defined linkage between exposure and effects in humans. One way to approach this is to generate quantitative data on gene expression that can be mechanistically linked to a toxicant mode-of-action, and ultimately linked to specific adverse effects. In order to conduct these studies in humans, this will require the development of assays which can be applied to easily accessible biological samples. One possible solution is to use array technology to monitor the expression of thousands of genes in a single blood sample from the test subjects. The aim of this approach is to find changes in gene expression that not only indicate if the subject has been exposed to a particular class of chemical, but also if pre-clinical responses to the environmental stressor have been initiated. For this approach to be of value, gene expression in the readily available surrogate tissues must correlate with gene expression in target organs such as the testis and be predictive for adverse effects.

If correlation of gene expression in target and surrogate tissues can be reproducibly demonstrated with various classes of chemicals in response to chronic and acute exposures, then the use of surrogate tissues may prove useful as a non-invasive method for biomonitoring toxicant exposure. Any combination of surrogate and target tissues can potentially be used, although it may be advantageous to pair more compatible tissues where possible. For example, blood cells may be more informative of the status of the thyroid or spleen than reproductive organs, and sperm may be a better indicator for the testis. The use of a surrogate tissue to monitor gene expression changes in tissues or organs elsewhere in the body could ultimately pave the way for early diagnosis of environmental exposures, offer prognostic indicators of toxicant exposure and health outcome, and permit the development of lifelong health monitoring programs.

Conclusion

One of the clearest windows into an organism's response to the environment is through its genome. Over the coming years, one of the great challenges facing toxicologists will be to harness the power of genomics, and the data it produces, to the goals of hazard identification, risk assessment, and protecting public health. This effort will be easier if it is mounted on a collaborative, international scale that brings to bear many resources and talents. For while the window is indeed very clear, making sense of all that we can see is a formidable task.

Acknowledgements

The author is grateful for the opportunity to work with all of the colleagues who contributed to this report, the names of whom can be found as co-authors in the references cited. The information in this document has been funded wholly (or in part) by the U.S. Environmental Protection Agency. It has been subjected to review by the National Health and Environmental Effects Research Laboratory and approved for publication. Approval does not signify that the contents reflect the views of the Agency, nor does mention of trade names or commercial products constitute endorsement or recommendation for use.

References

Rockett JC, Dix DJ (1999) Application of DNA arrays to toxicology. Environmental Health Perspectives 107:681-685

Rockett JC, Dix DJ (2000) DNA arrays: technology, options and toxicological applications. Xenobiotica 30:155-177

Rockett JC, Mapp FL, Garges JB, Luft JC, Mori C, Dix DJ (2001a) Effects of hyperthermia on spermatogenesis, apoptosis, gene expression, and fertility in adult male mice. Biology of Reproduction 65:229-239

Rockett JC, Luft JC, Garges JB, Krawetz SA, Hughes MR, Kim KH, Oudes AJ, Dix DJ (2001b) Development of a 950-gene DNA array for examining gene expression patterns in mouse testis. Genome Biology 2:research0014.1-0014.9

Future Prospects and Risk Assessment

Toxicology and Genetic Toxicology in the New Era of “Toxicogenomics”: Impact of "-omics" Technologies

Marilyn J. Aardema[1] and James T. MacGregor[2]

[1] The Procter & Gamble Co. Miami Valley Laboratories, P.O. Box 538707, Cincinnati OH 45252.
[2]U.S. Food and Drug Administration, National Center for Toxicological Research, 5600 Fishers Lane, Rockville, MD 20857.

The unprecedented advances in molecular biology during the last two decades have resulted in a dramatic increase in knowledge about gene structure and function, an immense database of genetic sequence information, and an impressive set of efficient new technologies for monitoring genetic sequences, genetic variation, and global functional gene expression. These advances have led to a new sub-discipline of toxicology: "toxicogenomics". We define toxicogenomics as “the study of the relationship between the structure and activity of the genome (the cellular complement of genes) and the adverse biological effects of exogenous agents.” This broad definition encompasses most of the variations in the current usage of this term, and in its broadest sense includes studies of the cellular products controlled by the genome (messenger RNAs, proteins, metabolites, etc.). The new “global” methods of measuring families of cellular molecules, such as RNA, proteins, and intermediary metabolites have been termed "-omic" technologies, based on their ability to characterize all, or most, members of a family of molecules in a single analysis. With these new tools, we can now obtain complete assessments of the functional activity of biochemical pathways, and of the structural genetic (sequence) differences among individuals and species, that were previously unattainable. These powerful new methods of high-throughput and multi-endpoint analysis, include gene expression arrays that will soon permit the simultaneous measurement of the expression of all human genes on a single "chip". Likewise, there are powerful new methods for protein analysis (proteomics: the study of the complement of proteins in the cell) and for analysis of cellular small molecules (metabonomics: the study of the cellular

---

This article has been reproduced from Mutation Research, Vol 499, 2002, pp 13-25, Aardema & MacGregor, by the permission of Elsevier Science, Ltd.

metabolites formed and degraded under genetic control). This will likely be extended in the near future to other important classes of biomolecules such as lipids, carbohydrates, etc. These assays provide a general capability for global assessment of many classes of cellular molecules, providing new approaches to assessing functional cellular alterations. These new methods have already facilitated significant advances in our understanding of the molecular responses to cell and tissue damage, and of perturbations in functional cellular systems.

As a result of this rapidly changing scientific environment, regulatory and industrial toxicology practice is poised to undergo dramatic change during the next decade. These advances present exciting opportunities for improved methods of identifying and evaluating potential human and environmental toxicants, and of monitoring the effects of exposures to these toxicants. These advances also present distinct challenges. For example, the significance of specific changes and the performance characteristics of new methods must be fully understood to avoid misinterpretation of data that could lead to inappropriate conclusions about the toxicity of a chemical or a mechanism of action. We discuss the likely impact of these advances on the fields of general and genetic toxicology, and risk assessment. We anticipate that these new technologies will 1) lead to new families of biomarkers that permit characterization and efficient monitoring of cellular perturbations, 2) provide an increased understanding of the influence of genetic variation on toxicological outcomes, and 3) allow definition of environmental causes of genetic alterations and their relationship to human disease. The broad application of these new approaches will likely erase the current distinctions among the fields of toxicology, pathology, genetic toxicology, and molecular genetics. Instead, a new integrated approach will likely emerge that involves a comprehensive understanding of genetic control of cellular functions, and of cellular responses to alterations in normal molecular structure and function.

Key words. Genetic toxicology, toxicogenomics, -omics, proteomics, metabonomics.

1. **Introduction**

The advances in biotechnology that have been achieved in the past few decades are truly extraordinary. They have provided an understanding of biological processes and an array of molecular technologies that have dramatically impacted all fields of biological science. This knowledge, and the technologies upon which it is based, have already impacted the fields of genetic and general toxicology, and provide the opportunity for further major advances in these fields. The current generation of scientists has already witnessed an extraordinarily rapid progression of knowledge of cellular and molecular genetics_from the discovery in 1944 that DNA was the genetic "blueprint" of life [1], to elucidation of the molecular structure of DNA [2], unraveling of the genetic code [3], and establishment of methods to efficiently "engineer" the genetic information in organisms [4-5]. The complete genetic sequence of the human genome has now been determined [6], and the sequences of many other important species are now nearing completion. Even though the pace at which these accomplishments were achieved is extraordinary, the future promises even more dramatic change.

Among the advances in technology that have been achieved, methods that allow global analysis of cellular constituents are of particular importance to the fields of general and genetic toxicology. Of these global methods, nucleic acid microarrays are currently receiving the greatest attention by the toxicology community. These technologies have made it possible to monitor thousands of nucleic acid sequences simultaneously—either specific expressed RNAs or sequence variants (polymorphisms) in DNA [7-8]. The analysis of expressed mRNAs using microarrays is analogous to conducting thousands of Northern blot analyses concurrently, and offers the possibility to monitor expression of individual genes across the entire genome (i.e. global gene expression analysis, or transcriptomics). Thus, microarray technology provides a means of studying multiple pathways and mechanisms at the same time. Such a global analysis of gene expression has the potential to provide a more comprehensive view of toxicity than has been possible previously, since toxicity generally involves change not only in a single or few genes but rather is a cascade of gene interactions. This unbiased form of analysis is certain to result in a more comprehensive picture of toxicological mechanisms and to lead to the re-evaluation of many of our currently-held beliefs.

Although nucleic acid microarray technologies have received much attention recently, other powerful new tools for global analysis of cellular constituents are already available and will also have a major impact on the field of toxicology. These include technologies for global analysis of proteins and peptides (proteomics), and of cellular metabolites (metabonomics). Among these advances are improvements in classical 2-dimensional (2D) gel electrophoresis, the introduction of multidimensional liquid chromatography, tandem mass spectrometry, and database searching technologies (termed multidimensional protein identification technology, or MudPIT), and improved mass spectroscopic identification of protein sequences using matrix- or surface-enhanced laser desorption ionization (MALDI, SELDI) techniques that allow rapid characterization of proteins or protein fragments [9-13]. These proteomic methods allow for the analysis of the functional and structural proteins in a sample. Methods for simultaneously monitoring small molecules involved in intermediary metabolic pathways (metabonomics) are also at hand [14-16] The ability to monitor defense responses via proteomics or metabonomics in humans at sub-pathological doses is of particular importance because it will make possible human studies that could not be carried out at overtly toxic exposures. These technologies provide complementary information to gene expression data. Clearly, post-translational modifications of proteins, such as phosphorylation, will not be evident as changes in gene expression. Also, nucleic acids may not be available for analysis in all cases (e.g., invasive procedures would be needed to obtain samples from many human tissues), although proteins may be secreted or diffuse into accessible compartments or be more amenable to imaging techniques. The use of all of these tools will be important for obtaining a comprehensive picture of toxicological changes in cellular constituents.

Thus, we now have available new families of technologies that provide a comprehensive analysis of the complete, or near-complete, cellular complement of specific constituents, such as RNAs, DNAs, proteins, intermediary metabolites, etc. These have been termed "-omics" technologies, a terminology derived from the Latin suffix "-ome" which denotes a body or group—in the commonly-used sense of a complete body or group (*e.g.*, the "biome"—the complement of living organisms in a particular environment, or the "genome"—the complete set of genes contained in the cellular complement of chromosomes) [17] . These now include genomics, transcriptomics, proteomics, and metabonomics. In the near future we may expect extension of these technologies to include other classes of cellular molecules, such as lipids, carbohydrates, lipoproteins, etc. These technologies are extremely powerful new tools with which to study disturbances of cellular homeostasis or structural integrity at a molecular level.

Although the molecular technologies for studying the genetic basis of disease, responses to molecular damage, and variation in response to toxicants have advanced in parallel with the science of cellular genetics, the regulatory approach to safety evaluation of new products has undergone little change over the past few decades. For example, the strategy for evaluating cell and tissue damage during product development described by Lehman et al. [18-19] has changed little during the past 50 years. Likewise, the strategy of genotoxicity testing remains similar to that developed during the 1970s [20]. On the other hand, carcinogenicity testing and risk assessment have recently begun to incorporate new mechanistic knowledge of the genetic basis of cancer induction, and biotechnology is being employed to construct animal models with specific genetic characteristics relevant to human carcinogenesis [21-23]. Two examples of the impact on regulatory science are the 1996 EPA draft carcinogenicity risk assessment guidelines, which ask that a mechanistic approach be taken to cancer risk assessment [24], and the introduction of specific transgenic animal models for the evaluation of cancer risk from exposure to pharmaceuticals [25, 22]. Thus, the potential of the powerful new tools that have facilitated the biotechnology revolution has just begun to be realized in the fields of regulatory toxicology, genetic toxicology, and carcinogenesis.

There is every reason to expect major change during the next decade, as new technologies and knowledge become incorporated into regulatory and industrial practice. Indeed, a new sub-discipline of "toxicogenomics" has already been recognized. We define toxicogenomics broadly—as the study of the relationship between the structure and activity of the genome (the cellular complement of genes) and the adverse biological effects of exogenous agents. This is consistent with the broad definition of pharmacogenomics recently proposed by Lesko and Woodcock [26]. Unlike other new approaches or methods in toxicology that have been adopted slowly, genomic, proteomic and metabonomic methods are being evaluated and adopted rapidly by industry, academia and regulatory agencies. This is evidence that the practice of toxicology has begun to change and that change can be expected to occur rapidly. As with most new toxicological methods and approaches, collaboration will be required to develop the data and approaches necessary to achieve worldwide acceptance and use. The implications for all involved in this field are profound. In the future, the marriage of the field of toxicology with genetics and molecular biologic techniques will likely lead to the eradication of sub-disciplines such as "genetic toxicology", "classical toxicology", or "toxicological pathology". Rather, the broad impact and application of the field of molecular genetics will transform the field of toxicology into a science based on molecular biochemical knowledge and techniques. Further, the separation between toxicology measures and efficacy measures may no longer exist—as many of the same molecular endpoints will be used to assess both the efficacy and toxicity of new drugs and chemicals. In this article, we summarize the opportunities and challenges created by these impending changes, with special emphasis on genetic toxicology, regulatory science, and product development.

2.0 Monitoring of Gene Expression and Alterations in Cellular Constituents in Response to Toxicants

The key concepts that underlie the application of new technologies for global analysis of cellular constituents to the field of toxicology originated with the recognition that application of these technologies could provide a more comprehensive view of cellular and molecular damage than was previously possible.

- It has been recognized that "...gene expression is altered either directly or indirectly as a result of toxicant exposure in almost all cases examined." [27]. Indeed, disturbance of normal function or structure essentially defines toxicity; hence, the ability to simultaneously monitor biochemical homeostasis in a wide variety of pathways provides a comprehensive means of identifying and monitoring toxic effects.

- Expression of certain genes is required to achieve pathological outcomes. It has even been claimed that "there (are) no toxicologically relevant outcomes *in vitro* or *in vivo*, with the possible exception of rapid necrosis, that do not require differential gene expression" [28]. Although this may be slightly overstated, it is, in fact, clear that most pathological processes are active events achieved under genetic control. Expression analysis provides a very powerful means of monitoring these processes.

- Changes in gene expression associated with toxicity are often more sensitive and characteristic of the toxic response than currently employed endpoints of pathology. See, for example, the discussion by Nuwaysir et al [29]. An ability to monitor defense responses and pre-pathological compensatory responses to cellular damage is certain to result in useful new biomarkers of sub-pathological cellular damage. Such biomarkers will be of particular importance because they will make possible human studies that could not be carried out at overtly toxic exposures.

- It is thought that patterns of gene expression, or "fingerprints", will be identified and these could be used as biomarkers of exposure and as a method of identifying mechanisms of toxicity [30]. Gene expression chips from many different species that permit thousands of genes to be monitored simultaneously are already available [31-32]. The next generation of expression chips can be expected to allow all expressed human genes to be monitored simultaneously. Analogous chips that can support similar monitoring in laboratory animal species are also expected. This capability of "global" monitoring of essentially all expressed genes provides the opportunity to characterize the patterns of gene expression associated with specific types of damage and/or specific classes of chemicals. Figure 1 illustrates these concepts.

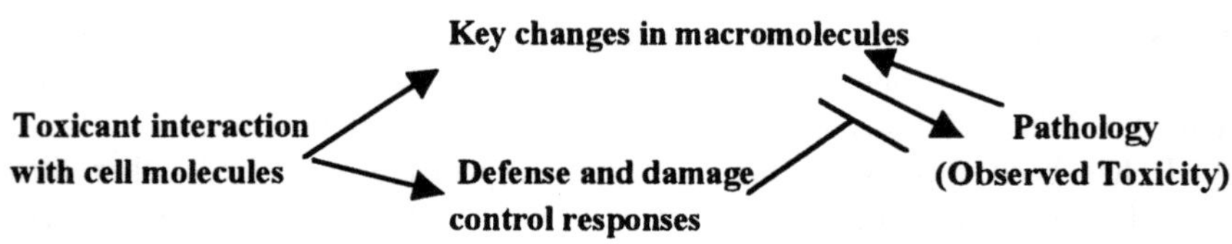

Figure 1. Basis of toxicogenomics

Observed Toxicity → **Characteristic profile of gene expression/ proteins/metabolites** → **Characterization and elucidation of mechanism**

Figure 2. Relationship between toxicity and development of fingerprints of gene/protein changes

3.0 Application of "-omic" Technoloiges-Promises and Pitfalls

The promises and challenges of some of the potential applications of global toxicogenomic approaches are described below. Certainly, uses beyond those mentioned here will develop as the field evolves.

3.1 Mode of action. Over the last several years, there has been an increased recognition by regulatory agencies of the value of mechanistic information for improving the risk assessment process. For instance, understanding the mode of action of a rodent carcinogen can help determine the relevance (or lack thereof) of carcinogenicity to humans or help in the quantitative extrapolation from high doses to low doses. The study of gene expression and genetically-controlled cellular constituents provides a new and powerful way of determining the mode of action. For example, association of a given toxic endpoint (e.g. carcinogenicity, genotoxicity, hepatoxicity) with a particular pattern of gene/protein expression, or a characteristic shift in cellular metabolites, may provide a "fingerprint" that is characteristic of a specific mechanism of induction of that toxicity (Figure 2). For example, recent studies by Samson's laboratory of yeast exposed to alkylating agents, oxidizing agents, and ionizing radiation demonstrated novel regulatory pathways dealing with the induced damage [33]. Exposure to alkylating agents was found to induce genes involved in DNA repair that were coregulated with genes involved with eliminating and replacing alkylated proteins [33-34]. This is an example of the type of information that can help determine which of the full profile of genes are involved in the mechanism of protecting cells against DNA damaging agents. Once a series of such fingerprints is defined for different mechanisms, the gene expression, protein, or metabolite pattern for a toxic chemical of unknown mechanism can then be compared to the established patterns for defined mechanisms. This provides a very powerful tool for categorization of toxicants according to mode of action. This is an important scientific advance, since defining the mechanism or mode of action of a chemical traditionally has required considerable time (often years), cost, and animal usage, and has included considerable uncertainty. Further, "fingerprints" of specific gene changes will provide the necessary information to design predictive toxicity screens (see below).

It may take years to build a database of well-characterized toxicities and related gene expression changes that is comprehensive enough to represent the many different modes of action of toxic agents. Because of the importance and the magnitude of the task of building comprehensive databases of expression profiles, collaboration among toxicologists in industry, regulatory agencies, and academic institutions is needed. Some important collaborations have already begun, such as the International Life Sciences Institute-Health and Environmental Sciences Institute (ISLI—HESI) project on the application of genomics and proteomics to mechanism-based risk assessment (http://www.ilsi.org) and the activities of the National Center for Toxicogenomics (www.niehs.nih.gov/nct/). As discussed previously, this sharing of data and experience is expected to stimulate a rapid evolution of the field of toxicology.

3.2 Predictive toxicology. In addition to the capacity to define mechanisms, global analysis of gene/RNA/protein/metabolite patterns offers the potential to predict toxic responses (figure 3). Since changes in these cellular molecules are thought to precede toxic outcomes, appropriate changes may serve as early, sensitive indicators of a potential toxicity. As discussed above, it is expected that similar chemicals, as defined by mechanism of toxicity (*e.g.*, DNA alkylating agents), will induce characteristic gene/protein expression patterns. Two recent studies [35-36] demonstrate this promise. In studies with genotoxic and non-genotoxic hepatoxins, patterns of gene expression changes were identified that may serve to distinguish chemicals with different mechanisms of action. Thus, it is expected that the profile of gene/protein expression changes will allow the discrimination among mechanisms of toxic damage—*e.g.*, between a DNA alkylating agent, a microtubule inhibitor, and an agent that perturbs protein structure. These unique patterns may then lead to the development of high through-put screening tests for use in predicting specific toxicities and also molecular biomarkers, applicable *in vivo*, that signal particular toxic mechanisms. For some chemicals, the results of predictive studies may provide sufficient information to be able to eliminate some conventional tests, or may provide a basis for the design of an appropriate bioassay with relevant species. Because these predictive tools have the potential to provide considerable savings in terms of time, cost, and animal use relative to traditional methods, it is expected that commercial promise will help drive rapid development of practical assays.

Key gene/protein/ metabolite alterations → **Characteristic expression "fingerprint"** → **Predicted toxicity associated with fingerprint**

Figure 3. Use of fingerprints of gene/protein expression changes to predict toxicity

The potential value of these new approaches to safety assessment has generated considerable excitement in the regulatory and industrial toxicology communities. However, it will be necessary to characterize multiple classes of agents with well-defined mechanisms of action before expression profiles for new biomarkers can be used reliably in regulatory decision-making [37]. It is critical that toxicologists in industry, regulatory agencies, and academic institutions develop a consensus, based on rigorous experimental data, about the reliability and interpretation of endpoints such as global gene expression patterns prior to use in regulatory and industrial settings.

3.3 Dose Response. A key aspect of the quantitative risk assessment process is understanding the effect of a chemical exposure at low doses typical of human exposure. Toxicity studies have traditionally been conducted at high doses, with various extrapolation methods used to estimate the effect at low doses. Quantitative differences in biochemical pathways and receptor affinities among species, coupled with the lack of biomarkers that respond to sub-pathological doses, has made it necessary to use exaggerated doses in order to minimize the chance of missing a toxic effect to which a particular species might be more resistant than is the human. Chemically-induced changes in gene/protein expression are expected to occur at doses of a chemical below those causing pathological alterations. Thus, by measuring gene/protein/metabolite changes at low doses, global analysis has the potential to provide critical information regarding biological effects at doses below those required for induction of the toxic endpoint. This should lead to improved extrapolations in the low dose range and a better determination of the effect of a chemical exposure to humans at low doses. It should also facilitate identification of threshold concentrations below which there is no significant risk.

A major challenge, of course, is to determine which of the molecular events that may change at low doses are necessary for pathological outcomes, *vs.* those that are adaptive, beneficial, and/or unrelated to the development of pathologies. Molecular genetics has revealed the molecular basis of key functional systems within the cell, and it has also demonstrated the presence of defense mechanisms to protect these key functions (*e.g.*, cell cycle control, structural integrity of proteins, control of free radicals associated with metabolism, and DNA repair mechanisms). The fact that these defenses have been highly conserved in evolution attests to their importance in limiting damage to these systems.

A specific example may serve to illustrate the various levels of understanding that are needed: the use of molecular responses to DNA damage as biomarkers of genetic toxicity. Maintenance of the fidelity of DNA replication is a prime example of the co-evolution of functional and defensive systems. The basic replicative DNA polymerase has co-evolved a proofreading function that corrects mispairings that occur during

DNA replication. Additionally, specific damage-recognition and repair molecules have evolved to protect the integrity of DNA and insure the fidelity of its replication, reviewed in [38], and more than 125 genes directly involved in DNA repair have already been identified in humans [39]. These include pathways to prevent highly damaged cells from replicating under conditions that would induce extreme damage, such as p53 and associated pathways. Thus, induction of genes associated with DNA repair is itself highly beneficial to the organism but is also a signal of the occurrence of DNA damage. If damage is extensive, then other responses may be induced, such as expression of p53 and GADD-mediated processes that prevent replication of highly damaged cells. Ultimately, induction of the caspase cascade that leads to cell death in order to eliminate highly genetically aberrant cells may occur. The relationships between these defense systems, and their role in responding to various types and levels of DNA damage must be understood in order to appropriately interpret these responses in the context of safety assessment and product regulation.

Analogous to the above example of responses to DNA damage, the functional molecules that control protein folding and export, and protein destruction required for cell cycle control and tissue remodeling, antigen processing, and other functions, have co-evolved to respond to protein damage. Similar to their normal physiological roles, these molecules can function as molecular chaperones or proteasomes to refold or destroy structurally damaged proteins [40-41]. Likewise, in the case of cellular energetics, defense systems have evolved to scavenge potentially toxic oxidative by-products and respond to perturbations that increase oxidative species within cells. Now that global analyses are becoming practical, it may be expected that relationships among different classes of damage will be found. For example, protein damage has already been associated with agents and conditions that cause DNA damage. It is not surprising that DNA-damaging agents would also damage RNAs, and thereby cause transcriptional errors. Such an association between DNA and protein damage has been noted in a number of studies. For example, in a recent study of gene expression changes induced by DNA damaging agents, Jelinsky et al [33] noted changes in genes involved in protein degradation and Radman and colleagues have noted that protein damage occurs both from transcriptional errors and also from protein structural alterations that lead to protein oxidation [42-43].

The biological relevance of gene expression changes, or of changes in functional levels of damage-response molecules, must be defined with respect to whether the gene(s) is pivotal in the pathway for a toxicity, if the change is reversible, and if the change in gene expression leads to altered cell or tissue function. It is important to guard against the temptation to classify every gene/protein expression change as adverse, as many changes will be physiological or reversible. Over-interpretation in the context of regulatory decision-making would result in inaccurate risk assessments and decision-making.

3.4 Species extrapolation. Because toxicity studies must often be conducted in animal models, extrapolation from results in animals to humans is needed to determine the potential human risk to a chemical exposure. The greatest single limitation of modern toxicological practice is probably the uncertainty of extrapolating from laboratory models to the human. Although the similarities in biochemistry and molecular biology among species has permitted a wide variety of useful laboratory models for the study of toxicological effects, quantitative differences often leave much uncertainty about differences in quantitative exposure-response relationships in the human compared with laboratory model systems. In almost all cases, quantitative differences in dose-response relationships exist between humans and model species, and in extreme cases biological responses to a given exposure may differ qualitatively. Thus, one of the great needs in the field is for "bridging biomarkers" of damage that can be used to compare toxic responses among species. In particular, human biomarkers that indicate that a given pathological condition is being approached—before it actually becomes manifest—are needed, so that it is ethically permissible to conduct human studies.

The increased understanding of cellular control and defense mechanisms, coupled with the powerful "-omic" technologies, have the potential to provide new bridging biomarkers that could greatly reduce this source of uncertainty in the risk assessment process (figure 4). Measurement of molecular events across a range of doses can be used to compare responses across species, including humans. The degree of similarity in gene expression pattern between different species will provide a new tool to help determine whether results in one experimental species are relevant to another—*i.e.*, a high degree of similarity in key genes in the gene expression fingerprint for a given exposure to a particular toxicant in a laboratory animal model compared to the expression pattern in a human (or human cells) should indicate that the molecular damage and response occurring in the animal model is similar to that in humans, and therefore relevant to anticipated human health outcomes. Likewise, dissimilar patterns could provide support for the lack of relevance of a unique animal toxicity finding to human risk. A similarity in gene expression patterns could also be used to select the most appropriate animal model prior to the conduct of toxicity studies. Understanding critical gene/protein/metabolite changes will also be useful for extrapolating from *in vitro* test systems to animal models.

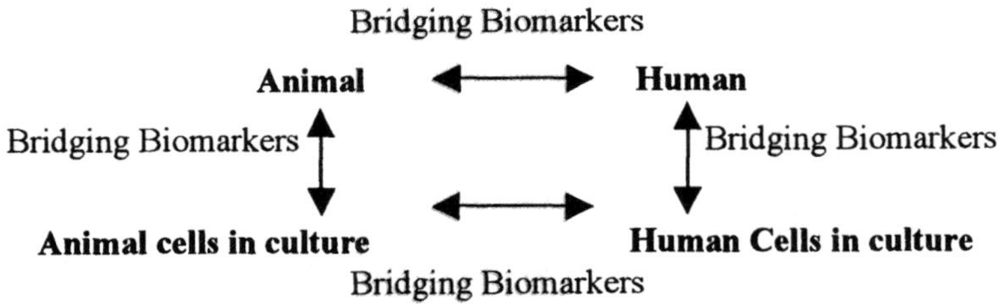

Figure 4. Use of Bridging Biomarkers To Extrapolate from Laboratory Models to Humans

3.5 Human biomarkers of exposure and effect. Another key component of the risk assessment process is establishing the extent of human exposure to a hazardous chemical or drug. Biomarkers of change in molecular genes/proteins/metabolites, should allow assessment of response to exposure, as well as loss of cellular integrity or disturbance of homeostasis in cells or tissues. These new biomarkers will be identified and validated based on our increasing knowledge of the biochemistry and genetics of molecular damage response, host-defense cell signaling, and cell death, and will have to take into account the pathway from protection and defense to pathological changes.

To identify systematically and validate biomarkers of effects, two distinct strategies will likely prove useful. First, effects of specific well-characterized pathologies on gene/protein/small molecules within the cell will need to be characterized to determine the relationship between these potential markers and specific types of damage. Second, "fingerprints" of cellular responses to classes of chemicals with known common biological effects will need to be investigated to develop a "library" of chemical class-specific cellular perturbations. These two strategies should lead to an understanding of changes in gene products, proteins, and specific small molecules that are characteristic of specific types of pathology, and thus to a new system of biological classification of chemicals based on similarities in their mechanisms of interaction with key cell receptors and response elements. Such studies will need to be conducted in important model species, beginning with well-established laboratory animal and cellular models, and ultimately undertaken in the human (such as during clinical trials) to understand differences and similarities in such responses among species. These biomarkers should allow monitoring and characterization of pathological damage as well as grouping of new chemicals into categories based on known responses to previously-studied chemicals.

3.6. Effect of multiple chemical exposures. Humans are rarely exposed to single chemicals at any given time. Most commercial products are a combination of chemicals. Nonetheless, because of practical technical limitations, most studies of toxicity evaluate one chemical at a time and risk assessments are traditionally conducted on single chemicals. It is virtually impossible to extrapolate results from a series of tests on individual chemicals to the effect of exposure to a complex mixture. Using high through-put genetic and proteomic approaches to measure changes in gene/protein expression after exposure to mixtures of toxicants has the potential to allow for the assessment of interactions such as additivity, synergism, or antagonism. Additionally, comparing gene expression patterns of test substances containing contaminants to a library of patterns for known genotoxins could help identify the presence of minor contaminants in products, eliminating the often tedious analytical approaches currently used to identify toxic contaminants.

4.0. Influence of Gene Sequence Variation (Polymorphisms) on Responses to Toxicants

In addition to the study of gene expression/proteins/metabolites in response to exposure to toxicants, new methods of evaluating gene sequence variation (polymorphisms) are also available. Such methods are making possible a systematic evaluation of the effects of variant genetic sequences on responses to toxicants. Examples of genetic variants that affect sensitivity to adverse effects from chemical exposures have long been known [44-48]. However, one of the major findings of the human genome project is that variation in base sequence between individuals averages approximately 1 in every 1000 base pairs (bp) [6]. Thus, sequence variants among individuals will be very common in essentially all genes. Further, these genetic variants have been shown to be important in many human diseases [49-51]. Understanding the relationship of this genetic variability to human disease and sensitivity to chemical exposure will greatly facilitate individual health risk assessments and extrapolation of findings from laboratory models to human risk [52].

"Mapping" of sequence differences between individuals with and without diseases (most often single base differences known as "single nucleotide polymorphisms", or "SNPs") is now revealing a growing number of disease susceptibility genes, as well as polymorphisms that determine individual diversity in drug responses [53]. As the tools now available are applied to the field of genetic toxicology, two major advances can be anticipated: 1) identification of polymorphisms responsible for sensitivity to toxicity from particular agents, and 2) identification of chemical-induced genetic changes associated with particular diseases.

4.1 Examples of Polymorphisms Associated with Susceptibility

Present examples of polymorphisms associated with individual susceptibility to chemical exposures consist mainly of metabolic polymorphisms that affect either activation or detoxification of particular toxicants [54-55]. A classical example is the genetically-determined rapid vs slow acetylator phenotype [56-57]. This polymorphism is responsible for the differences in sensitivity to isoniazide among sub-populations with these N-acetylase variants. Another example of the influence of genetic variation and toxicity is the sensitivity to fava bean toxicity among Mediterranean populations with glucose-6-phosphate dehydrogenase deficiency (G6PD), reviewed in [57]. Because of the very high frequency of base pair variation in the human genome, it is to be expected that genetic variability among individuals will occur at most, if not all, molecular targets for toxicants. It is also to be expected that this genetic variation will be found to be a major cause, or perhaps the major cause, for variation in susceptibility to toxicant exposure. This concept is supported by the exponentially growing list of spontaneous pathologies (diseases) associated with genetic variants, as many such variants would be expected to affect both spontaneous and chemically-induced pathologies. The knowledge that essentially every gene will vary in base sequence among individuals suggests that many examples of genetic determinants of variability in receptor conformation, in biochemical pathways, and in key defense molecules will be shown to be responsible for significant variations in sensitivity to different agents. Indeed, examples are already known—such as the polymorphisms in cardiac potassium channels that confer susceptibility to Q-T interval prolongation and susceptibility to fatal arrhythmias [53, 58].

As the genetic basis of various diseases becomes better understood, it is expected that the contribution of chemically-induced genetic change to those diseases will likewise be elucidated. Such correlations are presently best understood in the case of carcinogenesis, where the genetic basis of cancer is finally becoming understood and the relationship between chemically-induced genetic alterations and the development of carcinogenesis is well-established. The powerful genetic tools now available have made it possible to now consider studies of the relationship between chemical exposures and genetic predisposition to other diseases, such as aging, cardiovascular disease, and heritable biochemical diseases.

4.2. Identification of Polymorphisms

Initially, practical applications of methods for identifying effects of genetic variation among individuals on responses to chemical agents may be expected for those classes of agents with mechanisms of action that are well characterized, such as pharmaceuticals. Interest in applying these technologies to pharmaceutical development has been stimulated by the occurrence of idiosyncratic toxic responses observed in humans, often after a drug is introduced into the marketplace. Many of these idiosyncratic reactions are likely to be due to genetic variations among individual patients. Sensitivities to adverse effects from many drugs are already known to result from modification of drug metabolism pathways (*e.g.*, terfenidine, isoniazide [59-61]). As a result, gene arrays (chips) containing known variants of human enzymes for drug metabolism are being developed to identify these individuals (e.g. affymetrix CYP450 array, see www.affymetrix.com; eSensor™, see www.motorola.com/lifesciences). As gene-chip technology becomes more available, it is not far-fetched to expect that an individual's metabolic enzyme genotype might become a routine part of one's medical record, and be used in selecting appropriate drugs and adjusting doses on an individual basis.

These capabilities will also bring new ethical dilemmas to the field. A discussion of the ethical considerations of the application of genomics information to humans is beyond the scope of this paper, and is only mentioned here briefly to help illustrate the dilemma. There has been a question of whether individuals with polymorphisms known to sensitize them to cardiac Q-T prolongation should be actively sought and included in clinical trials in order to better understand the potential for unexpected adverse reactions in the general population, or whether such individuals should be actively excluded from clinical trials in order to minimize the risk to those in the trials at the expense of allowing the potential for adverse effects in the general population to remain relatively poorly defined will inevitably arise. Ethically sound strategies that protect individual subjects but also allow the generation of scientific knowledge needed to protect the population at large need to be developed in conjunction with the regulatory application of new genetic knowledge and technologies.

4.3 Laboratory Models with Human Genetic Characteristics

Our increased understanding of the relationship between the nucleotide sequences of genes and the cellular functions that they control will certainly allow better interpretation of the significance to humans of laboratory findings in animal and cellular models. In addition to allowing a better understanding of laboratory findings, advances in the field of genetics have now made it possible to manipulate the genetic structure of laboratory models through transgenic and other technologies. These technologies have made it practical to construct laboratory animal and cellular models that contain specific human genetic characteristics of interest. Thus, it is now possible to construct "humanized" models that contain either normal or polymorphic forms of human genes and cellular macromolecules. As important human targets for drug and toxicant interactions are identified and characterized, we can now expect that analogous laboratory models that allow these interactions to be studied in animals, cells, and tissues will become available. Such "humanized" models have already been created, demonstrating the feasibility of this approach. Examples include animal models of sickle-cell disease [62], and cell lines engineered to express the human cytochrome P450 drug-metabolizing enzymes [63,64].

5.0. Other Challenges

In addition to the issues discussed above, there are a number of general challenges to this new field of toxicogenomics. Genomic, proteomic and metabonomic methods are being evaluated or adopted by all sectors of industry, academia, and regulatory agencies at an unprecedented rate. These technologies are not yet standardized, and many formats are being used in different laboratories. Many laboratories are developing their own custom microarrays, which are typically prepared with specific subsets of genes of interest to the investigator (*e.g.*, toxicity related genes). The proprietary nature of some of these custom arrays prevents an evaluation of inter-laboratory reproducibility. This rapid adoption of new technologies will undoubtedly result in some publications of preliminary data without the appropriate validation and interpretation. This creates the need for caution in interpretation of data, until the strengths and limitations of these new tools are evaluated through carefully conducted studies in multiple laboratories.

A major issue with array technology is quality control and characterization of analytical performance. To obtain meaningful data, it is necessary that the sequences on the arrays are correctly identified and annotated. Further, the arrays need to be manufactured to a standard that yields reproducible results. Reproducibility, sensitivity, and robustness must be determined and controlled for each of the "spots" on the arrays that comprise many thousands of individual assays, and the biological meaning of alterations in specific expression patterns must also be determined.

Another challenge associated with the new "-omic" technologies is that of acquiring, storing, and analyzing the extensive amount of data that is generated by these studies. There is a need for sophisticated methods of storing and analyzing data, as well as standardized analysis approaches and algorithms that facilitate comparison of data among laboratories.

Soon, we may expect complete gene expression arrays that include all genes of a species being studied. However, at the present time most arrays are still incomplete, and are available for only a limited number of species. Thus, initial data will not be complete and only those genes available on the array for a few species will be measured. Of course, this is changing rapidly as updated arrays containing a large fraction of the genome are becoming available. It is anticipated that entire genomes of the species of most interest (*e.g.*, human, mouse, rat, etc.) will be represented on the chips to appear in the near future.

Measurements of toxicity must be quantitative, because quantitative dose-response relationships are the basis of evaluating effects in "real-world" exposure situations and regulatory decision-making. Therefore, the use of these new methods of global analysis must also be quantitative. Thus, it is necessary first to assure that analytical results are quantitatively reliable, and second to determine the magnitude of changes in expression or cellular concentrations that are toxicologically relevant. The dynamic range of changes in individual genes, the reversibility of changes observed, and the role of particular changes in determining cellular pathologies are all important considerations. This, of course, requires an understanding of the kinetic and quantitative relationships among genetic pathways and their relevance to cell and tissue function. To avoid falsely indicting chemicals as toxic (or erroneously thinking they are safe), the scientific community must establish consensus on these key factors.

6.0 The Future

As tools that enable simultaneous analysis of multiple targets and pathways are applied, a more comprehensive classification of cellular perturbations that result from toxicity will emerge. Already, toxicologists are starting to look at toxicity through the wider lens of toxicogenomic methods that provides a more complete view of complex networks of gene/protein changes. This is analogous to the difference between attempting to assess neurotoxicity by monitoring the activity of a single neuron at an arbitrary site vs simultaneously monitoring the function of thousands of neurons in multiple neural networks throughout the body. It is likely that these new technologies will reveal that many currently held beliefs are incomplete or incorrect. However, these new technologies, and those yet to be developed, will undoubtedly lead us to a better understanding of toxicity.

It is almost certain that global measures of gene structure, gene products, and small biomolecules will become inexpensive, and thus routine, as current methods are refined and additional ones introduced. Rather than analyzing endpoints of toxicity separately (*e.g.*, genotoxicity, reproduction/teratology, allergenicity) and in isolation, relevant genes for all toxicities may eventually be formatted on an array to provide a far more comprehensive toxicity analysis. Further, toxicity and efficacy measures may be conducted concurrently, as the same genes and gene products are involved in both pharmacological and toxicological responses. This will result in comprehensive, integrated, assessments of products with intended beneficial properties, such as pharmaceuticals and nutrients.

The toxicologist of the future will require a multidisciplinary understanding of genetics, cellular signal transduction and control, and conventional pathology. Through the fusion of these disciplines, a major beneficial transformation of the field is to be expected. The future of toxicology and genetic toxicology lies in the merger of these disciplines and the judicious application of genetic techniques and biomarkers to provide a more holistic understanding of the effects of chemicals on cellular alterations.

References

[1] O.T. Avery, C.M. McLeod, M. McCarty, Studies on the chemical nature of the substance inducing transformation of Pneumococcal types, J. Exp. Med. 79 (1944) 137-158.

[2] J.D. Watson, F.H.C. Crick, Molecular structure of nucleic acids: A structure for deoxyribose nucleic acid, Nature 171 (1953) 737-738.

[3] J.F. Speyer, P. Lengyel, C. Basilio, A.H. Wahba, R.S. Gardner, S. Ochoa, Synthetic polynucleotides and the amino acid code, Cold Spring Harbor Symp. Quant. Biol. 28 (1963) 559-568.

[4] S.N. Cohen, The manipulation of genes, Sci, Am. 233 (1975) 25-33.

[5] S.N. Cohen, Gene manipulation, N. Engl. J. Med. 294 (1976) 883-889.

[6] J.C. Venter, M.D. Adams, E.W. Myers, P.W. Li, et al., The Sequence of the Human Genome, Science 291 (2001) 1304-1351.

[7] P.O. Brown, D. Botstein, Exploring the new world of the genome with DNA microarrays, Nat. Genet. 21, Suppl 1 (1999) 33-37.

[8] M. Beekman, N. Lakenberg, S.S. Cherny, P. de Knijff, C.C. Kluft, G.J. van Ommen, G.P. Vogler, R.R. Frants, D. I. Boomsma, P.E. Slagboom, A powerful and rapid approach to human genome scanning using small quantities of genomic DNA, Genet Res 77 (2001) 129-34.

[9] N.L. Anderson, J. Taylor, J.P. Hofmann, R. Esquer-Blasco, S. Swift, N.G. Anderson, Simultaneous measurement of hundreds of liver proteins: application in assessment of liver function, Toxicologic Pathology 24, (1996) 72-76.

[10] G. MacBeath and S.L. Schreiber, Printing proteins as microarrays for high-throughput function determination, Science 289 (2000) 1760-1763.

[11] J.R. Yates, III, Mass spectrometry and the age of the proteome, J. Mass Spectrom. 33 (1998) 1-19.

[12] J.R. Yates, III, Mass spectrometry. From genomics to proteomics, Trends Genet. 16 (2000) 5-8.

[13] M.P. Washburn, D. Wolters, J.R. Yates, III, Large-scale analysis of the yeast proteome by multidimensional protein identification technology, Nat. Biotechnol. 19 (2001) 242-247.

[14] J.K. Nicholson, J. C. Lindon, E. Homes, "Metabonomics": understanding the metabolic responses of living systems to pathophysiological stimuli via multivariate statistical analysis of biological NMR spectroscopic data, Xenobiotica 29 (1999) 1181-1189.

[15] E. Holmes, A.W. Nicholls, J.C. Lindon, S.C. Connor, J.C. Connelly, J.N. Haselden, S.J. Damment, M. Spraul, P. Neidig, J.K. Nicholson, Chemometric models for toxicity classification based on NMR spectra of biofluids, Chem Res Toxicol 13 (2000) 471-478.

[16] E. Holmes, J.K. Nicholson, G. Tranter, Metabonomic characterization of genetic variations in toxicological and metabolic responses using probabilistic neural networks. Chem. Res. Toxicol. 14 (2001) 182-191.

[17] J. Lederberg and A. T. McCray, "Ome sweet "Omics-a genealogical treasury of words, The Scientist 15 (2001) p.8.

[18] A.J. Lehman, E.P. Lang, G. Woodard, J.H. Draize, O.G. Fitzhugh, A.A. Nelson, Procedures for the appraisal of the toxicity of chemicals in foods, Food, Drug, Cosmetic Law Quarterly 4 (1949) 412-434.

[19] J.M. Barnes, F.A. Denz, Experimental methods used in determining chronic toxicity, Pharm. Rev. 6 (1954) 191-242.

[20] J.T. Macgregor, D.A. Casciano, L. Müller, Strategies and testing methods for identifying mutagenic risks. Mutation Res. 455 (2000) 3-21.

[21] T.L. Goldsworthy, L. Recio, K. Brown, L.A. Donehower, J.C. Mirsalis, R.W. Tennant, I.F.H. Purchase, Transgenic animals in toxicology, Fund. Appl. Toxicol. 22 (1994) 8-19.

[22] J. MacDonald, Evaluation of new models. II, in: P.F. D'Arcy, D.W.G. Harron (Eds.), *Proc. Fourth Int. Conf. On Harmonization, Brussels 1997,* Greystone Books Ltd., Antrim, N. Ireland, 1998, pp. 272-277.

[23] S.K. Sharan, A. Bradley, Role of transgenic mice in identification and characterization of tumor suppressor genes, Cancer Surv. 25 (1995) 143-159.

[24] U.S. Environmental Protection Agency, Guidelines for carcinogen risk assessment, Fed. Regulations 61 (1996) 17960-18011.

[25] J. DeGeorge, Carcinogenicity testing: A new approach, in: P.F. D'Arcy, D.W.G. Harron (Eds.), *Proc. Fourth Int. Conf. On Harmonization, Brussels 1997,* Greystone Books Ltd., Antrim, N. Ireland, 1998, pp. 261-263.

[26] L.J. Lesko, J. Woodcock, Pharmacogenomic-guided drug development: Regulatory perspective, Pharmacogen. J., in press.

[27] J.C Corton, S.P. Anderson, A.J. Stauber, D.B. Janszen, J.S. Kimbell, R.B. Conolly, Entering the era of toxicogenomics with DNA microarrays, in CIIT Activities 19, 2 (1999) 1-9.

[28] S. Farr, R.T. Dunn, II, Concise review: gene expression applied to toxicology, Toxicol. Sci. 50 (1999) 1-9.

[29] E.F. Nuwaysir, M. Bittner, J. Trent, J.C. Barrett, C.A. Afshari, Microarrays and toxicology: the advent of toxicogenomics, Mol. Carcinog. 24 (1999) 153-159.

[30] W.A. Pennie, Use of cDNA microarrays to probe and understand the toxicological consequences of altered gene expression, Tox. Letters 112-113 (2000) 473-477.

[31] J.C. Rockett, D.J. Dix, Application of DNA arrays to toxicology, Env. Health Perspectives 107, (1999) 681-685.

[32] J.E. Celias, M. Kruhffer, I. Gromova, C. Frederiksen, M. Østergaard, T. Thykjaer, P. Gromov, J. Yu, H. Pálsdóttir, N. Magnusson, T. F. Ørntoft, Gene Expression profiling: monitoring transcription and translation products using DNA microarrays and proteomics, FEBS Letters 480, (2000) 2-16.

[33] S.A. Jelinsky, P. Estep, G.M. Church, L.D. Samson, Regulatory networks revealed by transcriptional profiling of damaged Saccharomyces cerevisiae

cells: Rpn4 links base excision repair with proteosomes, Mole. Cell Biol. 20 (2000) 8157-8167.

[34] S.A. Jelinsky, L.D. Samson, Global response of Saccharomyces cerevisiae to an alkylating agent, Proc. Natl. Acad. Sci. 96 (1999) 1486-1491.

[35] M.E. Burzynski, M. McMillian, J. Ciervo, L. Li, J. B. Parker, R. T. Dunn, S. Hicken, S. Farr, M. D. Johnson, Toxicogenomics-based discrimination of toxic mechanisms in $HepG_2$ human hepatoma cells, Tox. Sci. 58 (2000) 399-415.

[36] J.F. Waring, R. Ciurlionis, R.A. Jolly,. M. Heindel, R.G. Ulrich, Microarray analysis of hepatotoxins in vitro reveals a correlation between gene expression profiles and mechanisms of toxicity, Tox. Letters 120 (2001) 359-368.

[37] J.C. Corton, A. J. Stauber, Toward construction of a transcript profile database predicitive of chemical toxicity, Tox. Sci. 58 (2000) 217-219.

[38] T. Lindahl, D. Wood, Quality control by DNA repair, Science 286 (1999) 1897-1905.

[39] A. Ronen, B.W. Glickman, Human DNA repair genes, Env. Molec. Mutagenesis 37 (2001) 241-283.

[40] L. Ellgaard, M. Molinari, A. Helenius, Setting the Standards:quality control in the secretory pathway, Science, 286 (1999) 1882-1888

[41] S. Wickner, M.R. Maurizi, S. Gottesman, Posttranslational quality control: folding , refolding and degrading proteins, Science 286 (1999) 1888-1893.

[42] S. Dukan, A. Farewell, M. Ballesteros, F. Taddei, M. Radman, T. Nyström, Protein oxidation in response to increased transcriptional or translational errors, Proc. Natl. Acad. Sci., USA 97 (2000) 5746-5749.

[43] F. Taddei, H. Hayakawa, M.-F. Bouton, A.-M. Cirinesi, I. Matic, M. Sekiguchi, M. Radman, Counteraction by MutT protein of transcriptional errors caused by oxidative damage, Science 278 (1997) 128-130.

[44] J.J. Burns, Pharmacogenetics and drug toxicity. Variation of drug metabolism in animals and prediction of drug action in man, Ann. N.Y. Acad. Sci. 151 (1968) 959-967.

[45] W. Kalow, Contribution of hereditary factors to the response to drugs, Fed. Proc. 24 (1965) 1259-1265.

[46] W. Kalow, Genetic aspects of drug safety, Appl. Ther. 8 (1966) 44-47.

[47] W. Kalow, Pharmacogenetics, Postgrad. Med. 42 (1967) 32-36.

[48] W. Kalow, Drug metabolism enzymes. Pharmacogenetics in animals and man, Ann. N.Y. Acad. Sci. 151 (1968) 694-698.

[49] K. Davies, After the genome: DNA and human disease, Cell 104 (2001) 465-467.

[50] F. Gilbert, Disease genes and chromosomes: disease maps of the human genome, Genet. Test. 4 (2000) 409-426.

[51] J. Ohashi, T.K. Tokunaga, The power of genome-wide association studies of complex disease genes: Statistical limitations of indirect approaches using SNP markers, J. Hum. Genet. 46 (2001) 478-482.

[52] M. Pirmohamed, B.K. Park, Genetic susceptibility to adverse drug reactions, Trends Pharmacol. Sci. 2001, 22 (2001) 298-305.

[53] W.W. Weber, Effect of pharmacogenetics on medicine, Environ. Molec. Mutagenesis 37 (2001) 179-184.

[54] K.R. Kaderlik, F.F. Kadlubar, Metabolic polymorphisms and carcinogen-DNA adduct formation in human populations, Pharmacogenetics 5 (1995) 108-117.

[55] D.W. Nebert, Drug-metabolizing enzymes, polymorphisms and interindividual response to environmental toxicants, Clin. Chem. Lab. Med. 38 (2000) 857-861.

[56] F.F. Kadlubar, Biochemical individuality and its implications for drug and carcinogen metabolism: recent insights from acetyltransferase and cytochrome P4501A2 phenotyping and genotyping in humans, Drug Metab. Res. 26 (1994) 37-46.

[57] W.W. Weber, Populations and genetic polymorphisms. Mol Diagn 4 (1999) 299-307.

[58] L.A. Larsen, P.S. Andersen, J. Kanters, I.H. Svendsen, J.R. Jacobsen, J. Vuust, G. Wettrell, L. Tranebjaerg, J. Bathen, M. Christiansen, Screening for mutations and polymorphisms in the genes KCNH2 and KCNE2 encoding the cardiac HERG/MiRP1 ion channel: implications for acquired and congenital long Q-T syndrome, Clin. Chem. 47 (2001) 1390-1395.

[59] J.A. Hey, M. delPrado, W. Kreutner, R.W. Egan, Cardiotoxic and drug interaction profile of the second generation antihistamines ebastine and terfenadine in an experimental animal model of torsade de pointes, Arzneimittel Forschung 46 (1996) 832-833.

[60] S.J. Smith, Cardiovascular toxicity of antihistamines, Otolaryngol. Head Neck Surg. 111 (1994) 348-354.

[61] A. Upton, N. Johnson, J. Sandy, E. Sim, Arylamine N-acetyltransferases - of mice, men and microorganisms, Trends Pharmacol Sci 22 (2001) 140-146.

[62] M.E. Fabry, Transgenic animal models of sickle cell disease, Experientia 49 (1993) 28-36

[63] C.L. Crespi, R. Langenbach, B.W. Penman, BW, Human cell lines, derived from AHH-1 TK +/- lymphoblasts, genetically engineered for expression of cytochromes P450, Toxicology 82 (1993) 89-104.

[64] C.L. Crespi, and B.W. Penman, Use of cDNA-expressed human cytochrome P450 enzymes to study potential drug-drug interactions. Adv Pharmacol 43 (1997)171-88.

Use of Genomics in Mechanism Based Risk Assessment

Denise E. Robinson[1], Syril D. Pettit[1], and D. Gwyn Morgan[2]

[1]ILSI Health and Environmental Sciences Institute, 1 Thomas Circle, NW, Ninth Floor, Washington, D.C., 20005-5802, USA; [2] 230 Windermere Ave; Wayne, PA 19087

Summary. Recent advances in genomic and proteomic research, coupled with the availability of novel tools and methods with which to analyze the products of altered gene expression, are providing new insights into the study of mechanistic toxicology. One of the short-term needs is to establish a body of publicly available knowledge to serve as a foundation for applying the data generated by these new methods to risk assessment. The ILSI Health and Environmental Sciences Institute is providing a scientific forum to facilitate further development and advances in genomics and proteomics methodologies toward the goal of increasing the utility of gene expression data for mechanism-based risk assessment. A collaborative research effort has been undertaken to help establish appropriate experimental methodologies for measuring alterations in gene expression and to relate genomic and proteomic expression data to toxic endpoints characterized by established methods under standardized conditions. Through this collaboration, data are being generated in government, industry and research laboratories on a series of prototype compounds across a broad range of available microarray platforms. This paper describes the initial phase of this project which has focused on nephrotoxicity, hepatotoxicity, and genotoxicity as important endpoints for the risk assessment of many chemicals. In addition, the challenges associated with drawing correlations between patterns of gene expression with biological and toxicological parameters, the influence of experimental conditions on data variability and reproducibility, and other issues associated with interpretation and application of data from these types of studies for risk assessment are identified.

Key words. Genomics, risk assessment, toxicity mechanisms, nephrotoxicity, genotoxicity, hepatotoxicity

Introduction

Recent advances in genomic and proteomic research, coupled with the availability of novel tools and methods with which to analyze altered gene expression and its transcriptional products, have the potential to provide new insights into mechanistic toxicology. These initiatives have given impetus to a new discipline, toxicogenomics, which has been defined as the application of information obtained using genomics-based methods to identify genes involved in toxicity and to better understand toxicological mechanisms, extending our scientific knowledge and contributing to more accurate human health risk assessments.

Knowledge of molecular mechanisms, which are the basis of toxicology, gained from well-designed experiments exploring the relationship between chemical exposure and altered gene and protein expression is key to establishing the relevance of animal data to humans. An improved understanding of the relevance of the toxicity manifested in currently available animal models will allow more meaningful extrapolation of these observations to humans and should facilitate identification of novel, clinically relevant biomarkers of toxicity or potential adverse effects. Identifying such early biomarkers of toxicity, both in vivo and in vitro, will enhance the development of new, more predictive short-term toxicity screens for chronic toxicities, including those for which there are no suitable screens currently available. Better integration will be possible between in vivo and in vitro assays. These developments should improve the currently available test methods for hazard identification and human health risk assessment, leading to the more rapid discovery and development of drug candidates with better toxicity profiles, thus improving the safety and quality of medicines and other therapeutic approaches in the future.

While these advances promise to revolutionize our ability to characterize hazard, there are a number of hurdles to be crossed before the potential of this rapidly evolving technology is fully realized. Although these technologies are commercially available and widely used, there is a lack of openly available information as well as comparable data from consistent study designs to analyze the suitability of individual approaches for various research and regulatory purposes. The ability to interpret the findings from a given experiment is complicated by the diversity of approaches used for data analysis. From a biological perspective, an understanding of how transcription products translate into predictors of toxicity is far from complete. This, coupled with a lack of consensus on determining the significance of gene expression data derived from different experimental protocols, compromises the ability to interpret the biological significance of such data as they relate to specific mechanisms of toxicity. Uniform standards of practice are needed, both with respect to study design and data evaluation. Furthermore, superimposed upon the decision to conduct these types of studies on new candidate compounds is the uncertainty related to how these data may be used in a regulatory setting.

While these uncertainties currently may dictate caution in the interpretation of data obtained by the use of these new methods, ongoing scientific and technological advances and their potential benefits indicate that there will be solutions. Much careful experimental work, data evaluation, and sharing of knowledge will be required to realize the full potential of these new methods. One short-term need is to provide a mechanism for developing scientific consensus around a body of publicly available knowledge that can serve as a foundation for applying the data generated by these new methods to assess risk.

The International Life Science Institute's (ILSI) Health and Environmental Sciences Institute (HESI), a non-profit research and educational organization based in Washington, DC, is providing an international forum for advancing the scientific basis for the development and application of genomic and proteomic methodologies in mechanism-based risk assessment. The ILSI HESI Committee on Application of Genomics to Mechanism-Based Risk Assessment has been organized as a consortium of scientists from industry, government and academic laboratories in the US, Europe and Japan. The committee is supported by the institutional members listed in Table 1. The Steering Committee and Scientific Advisors are listed in Tables 2 and 3, respectively. The committee's long-term objectives are to:

- provide a scientific forum to facilitate advances in genomics and proteomics,
- establish a multinational collaboration among scientists from government, industry, and academia,
- create publicly available data on gene and protein expression from toxicity studies,
- improve the understanding of molecular pathways of selected toxicities, and
- provide a rational basis for identification of appropriate biomarkers and interspecies extrapolation.

Short-term objectives have been established to begin addressing these long-term needs. A description of this collaborative research and evaluation program and the current status of the experimental work are described below, followed by a brief review of future goals and directions.

Table 1. ILSI HESI Genomics Committee – Participating Institutions

Abbott Laboratories	National Institute of Environmental Health Sciences
Amgen, Inc.	Novartis Pharmaceuticals Corporation
AstraZeneca	Pfizer, Inc.
Aventis Pharmaceuticals	Pharmacia Corporation
Bayer AG	The Procter & Gamble Company
Berlex Laboratories, Inc.	RIVM
Boehringer-Ingelheim Pharmaceuticals, Inc.	Sankyo Co., Ltd.
Bristol-Myers Squibb Co.	Sanofi-Synthelabo, Inc.
The Dow Chemical Company	Schering AG
Eisai Co., Ltd.	Schering-Plough Research Institute
Eli Lilly and Company	Tanabe Seiyaku Co., Ltd.
GlaxoSmithKline	U.S. Environmental Protection Agency/ Office of Research and Development
Hoffmann-La Roche, Inc.	U.S. Food and Drug Administration/ Center for Biologicals Evaluation and Research
Johnson & Johnson Pharmaceuticals	U.S. Food and Drug Administration/ Center for Drug Evaluation and Research
Medical College of Wisconsin	Unilever Research
Meiji Seika Kaisha, Ltd.	University of Surrey
Merck and Company	University of Wisconsin
Mitsubishi-Tokyo Pharmaceuticals, Inc.	Wyeth-Ayerst Research
National Center for Toxicological Research, U.S. FDA	Yamanouchi Pharmaceutical Co., Ltd.
National Institute of Health Sciences, Japan	
National Institutes of Health/ National Cancer Institute	

Table 2. ILSI HESI Genomics Committee Steering Committee

Dr. Marilyn Aardema	The Procter & Gamble Company
Dr. Christopher Bradfield	University of Wisconsin
Dr. Bruce Car	Bristol-Myers Squibb Co.
Dr. Mark Johnson	Johnson & Johnson Pharmaceuticals
Dr. Jeffrey Kramer	Pharmacia Corporation
Dr. Peter Lord	GlaxoSmithKline
Dr. William Mattes	Pharmacia Corporation
Dr. D. Gwyn Morgan	Retired
Dr. Ronald Newton	Eli Lilly and Company
Dr. William Pennie	Pfizer, Inc.
Ms. Syril Pettit	ILSI Health and Environmental Sciences Institute
Dr. Denise Robinson	ILSI Health and Environmental Sciences Institute
Dr. James Stevens	Eli Lilly and Company
Dr. Ronald Tyler	GlaxoSmithKline
Dr. Roger Ulrich	Rosetta Inpharmatics/Merck

Table 3. ILSI HESI Genomics Committee Scientific Advisors

Dr. Christopher Bradfield	University of Wisconsin
Dr. Albert Fornace	National Cancer Institute, NIH
Dr. G. Gordon Gibson	University of Surrey
Dr. Richard Roman	Medical College of Wisconsin

Experimental Design

The first of three short-term objectives of the ILSI HESI Genomics Committee has been to evaluate available experimental methodologies for measuring alterations in gene expression. To meet this objective, protocols were developed to evaluate the toxicity profiles of several prototypic genotoxicants, hepatotoxicants, and nephrotoxicants using both traditional toxicity measures as well as gene and protein expression techniques. These three toxicity endpoints were selected because they are relatively common target organs in acute, subchronic and chronic toxicity studies, because they present significant challenges in terms of risk assessment, or because increased understanding of the molecular mechanisms of toxicity would contribute significantly to the quality and reliability of the risk assessment. For each endpoint, several well-studied prototype compounds were chosen for study with the goal of comparing gene expression changes with biological measurements of toxicity in the context of the known toxicity profiles of each compound.

Protocols were developed to investigate temporal relationships and dose-response relationships of both gene expression changes and toxicity outcomes, as well as variability in the data generated within and across laboratories and across platforms. The overall study design incorporated measurements of standard indicators of toxicity, collection of RNA samples, and microarray-based gene expression analysis and protein evaluation by the various participating public and industry laboratories. Elements of the basic protocols were standardized whereas the microarray component was conducted according to each laboratory's standard operating procedures. Each participating laboratory volunteered to join one or more of the toxicity working groups and to contribute data using the microarrays and experimental protocols employed in their respective facilities. Microarray technologies from a broad range of commercial sources were used along with custom arrays designed and produced in the laboratories of participating organizations.

Nephrotoxicity protocols

Nephrotoxicity was chosen as an endpoint because it is a significant target organ that has caused regulatory agency concern and presents an ongoing risk assessment challenge for both new and approved medicines. Current methods to detect nephrotoxicity are relatively insensitive and manifestations of acute nephrotoxicity often are not useful in predicting longer-term outcomes. An understanding of the molecular mechanisms of toxicity would contribute to the identification of appropriate biomarkers, better selection of doses for toxicity studies and a more reliable extrapolation of risk associated with potential renal injury from experimental animal models to humans. The ILSI HESI Nephrotoxicity Working Group elected to focus on tubular, glomerular and interstitial toxicants. Three compounds were studied in male Sprague Dawley rats – cisplatin (0, 0.3, 1.0, and 5.0 mg/kg body weight i.p), gentamycin (0, 2, 10, 80 mg/kg body weight i.p.) and puromycin 0, 5, 20, 150 mg/kg body weight i.p.). Exposures ranged from 4 hours to 21 days, depending on the compound and the mechanism of toxicity.

Kidneys were sampled for RNA and processed for histopathology. Hematology, serum chemistry and urinalysis measurements were performed as well as protein analyses in plasma and urine. Both pooled RNA samples as well as individual animal

RNA samples are under analysis. PCR analysis has been utilized to confirm gene changes. So far, the genes that have shown increased or decreased expression appear to be consistent with known mechanisms of toxicity for these compounds. Effects have been detected as early as 4-hours post treatment. Detailed results will be published during 2002 by participating members and organizations.

Hepatotoxicity protocols

There was an early consensus to include hepatotoxicity as one of the areas of focus, because the liver is a frequent target of toxicity which often leads to regulatory action. In addition, mechanisms of hepatotoxicity have been relatively well-studied compared to other tissues. Methapyrilene and clofibrate were chosen as representative of distinct mechanisms of action with well-known hepatic toxicity profiles in male Sprague Dawley rats. Two doses at 24 hrs, 3 and 7 days were tested for each compound (0, 25, 250 mg/kg body weight per day for clofibrate; 0, 10, 100 mg/kg body weight per day for methapyrilene). Duplicate in-life studies were performed at 2 different sites to assess in-life variability across laboratories.

As with the nephrotoxicity experiments, livers and kidneys were saved for histopathology and RNA preparation, with remaining tissue frozen for future use. Clinical chemistry and pathology measurements demonstrated the expected toxicity profiles for the compounds. Both pooled RNA samples as well as individual animal RNA samples are under analysis on all microarray technologies volunteered for use in the ILSI HESI program. PCR analysis has been utilized to confirm gene changes. Evaluation of proteins in both serum and urine samples from study animals are underway. In addition, several laboratories volunteered to conduct parallel studies using primary rat hepatocytes to allow comparison of in vivo and in vitro studies. This set of studies will evaluate the similarities and differences across the various platforms with the data to be presented and published in 2002.

Genotoxicity protocols

Genotoxicity has long been a critical endpoint for risk assessment and usually provides some of the earliest data available on a compound. Positive genotoxicity data immediately raise concern regarding a compound's safety and generally lead to a linear risk assessment approach in regulatory settings where quantitative risk assessment is utilized. Where a weight of evidence approach is used, these data dictate a shift down a more conservative branch of the risk assessment decision tree, often terminating a compound's development or restricting its further use. This endpoint was selected because mechanisms of genotoxicity have been well-studied and a significant amount of data are available on a wide range of compounds. It should be possible to determine if the various forms of DNA and chromosomal damage detected by conventional assays possess a characteristic fingerprint discernable by characteristic changes in gene expression patterns. Such a correlation may be more likely to exist between gene expression and conventional genotoxic endpoints than might be the case for toxicity in complex organs and tissues. As such, this set of studies may hold the promise of providing more immediate insight into the applicability of these new methods for defining mechanisms to support a more precise risk assessment.

As a first set of studies, the genotoxicity working group picked 13 different acting mutagens and clastogens (cisplatin, benz[a]pyrene, MMS, bleomycin, hydrogen peroxide, taxol, N-AAAF, mitomycin C, 4-NQO, ENU, x-rays, etoposide, hydroxyurea. In addition, sodium chloride was tested at cytotoxic concentrations. These compounds were tested in one or more in vitro assay systems, including the mouse lymphoma cell assay and human TK6 cells. Cell cultures were exposed for 4 hours and immediately harvested or exposed for 4 hours followed by a 20-hour recovery period. Cytotoxicity was assessed as was mutagenicity by at least one of the following: micronucleus, TK gene mutation, and/or DNA adducts. All study components were performed in-house by each participating laboratory, including the cell culture, chemical treatments, mutagenicity measurements, isolation of RNA and gene expression analysis. Gene expression analysis and data evaluation are in progress.

Need for Development of Publicly Available Information

The second short-term objective of the ILSI HESI Genomics Committee is to contribute to the development of international databases linking gene expression data and key biological parameters. One of the key challenges identified at the outset of the ILSI HESI collaboration was the limited data from well-designed and well-controlled studies that were available in the public domain for joint evaluation by academic, government and industrial scientists. To work toward this objective, a short-term web-based gene expression data repository was built to house the results of the current studies of the ILSI HESI Genomics Committee. To standardize data submission to this repository from a variety of platforms, the Basic Expression Exchange Format (BEEF) was developed. The immediate goal was to provide a mechanism for sharing data among participants and to facilitate data analysis. In parallel with this effort and in keeping with the overall project goals, a collaboration with the European Bioinformatics Institute was initiated in late 2001. Through this collaboration, the ILSI HESI Genomics Committee will consider appropriate linkages between the diverse sets of experimental data, including biological data, gene expression data, microarray and experimental design information. One of the goals of this initiative is to provide a mechanism for public access to the committee-generated dataset as soon as possible following the completion of data analysis.

Evaluation of Current Experience and Future Issues

As the data from these carefully-designed and standardized studies emerge, it will be possible to assess the influence of several different variables on the reproducibility and variability of the experimental outcomes, particularly as they relate to differences between technology platforms, individual laboratories, experimental protocols, and the various approaches to data analysis and interpretation which determine the statistical as well as the biological significance of the results obtained. Thus far, the data are being evaluated primarily on a qualitative basis with most attention being given to drawing links between gene expression and the traditional toxicity endpoints to assess the biological relevance of the findings. In addition to the assessing the value of the various analytical methods available, it is also recognized that the appropriate application of statistical methods is an important future point of discussion.

While the data are still undergoing analysis, the interim findings have stimulated intense discussion and debate regarding "best practices" for experimental design, methodology, and data evaluation. Several key issues have been identified. As with any experiment, it is important to identify and reduce sources of experimental variability. While the initial expectation was that there would be significant overlap in the findings across the different platforms, it has been demonstrated that there are unique aspects to each technology. It is clear that there are substantial differences among the platforms and that use of different platforms enriches the overall data set. Another consideration is that different laboratories have developed individual approaches to data analysis, which can lead to discrepancies when making comparisons across laboratories. Thus, certain common metrics may need to be developed and that standardized approaches should be utilized in tandem with customized individual analyses.

The final analysis of the emerging data sets will contribute to the realization of the third objective of the ILSI HESI Genomics Committee – to determine whether known mechanisms and pathways of toxicity can be associated with characteristic gene expression profiles. This objective is held in high priority by a number of organizations including, but not limited to, those participating in this collaborative program. To be of utility, these new technologies must usefully and reliably characterize basic molecular mechanisms of toxicity in a meaningful and consistent fashion. Immediate and longer term needs are to understand and reduce variability across experiments and to consider a paradigm for the understanding and acceptance of new methods that takes into account the rapid evolution of the technology. It is the overarching objective of this project to continue to build scientific consensus on the significance of gene expression changes in the context of defining hazard and risk to human health, and to gain an understanding of the total body of evidence necessary to support sound scientific and regulatory decision making processes. Achieving this ultimate objective will require ongoing partnerships, public participation and continued consensus building efforts on an international scale.

Chemogenomics for Predictive Drug Assessment

Mike Furness, Cecelia Pearson, George Natsoulis, Alan Engelberg, Keith Bostian, and Kurt Jarnagin

Iconix Pharmaceuticals, 850 Maude Ave., Mountain View, CA 94043, USA

Summary. Improved lead prioritization in relation to toxicity, efficacy and mechanism of action could improve the overall success of drug discovery. DrugMatrix™ is an integrated database incorporating compound data including molecular pharmacology, gene expression, chemical structure and SAR, toxicology, pathology, and pharmacology. This database helps drug prioritization by providing the front line of drug discovery (toxicologists, pharmacologists and medicinal chemists) early data about adverse events and potential benefits. We have used DrugMatrix molecular pharmacology assays and expression findings to classify several nuclear hormone receptor agonists (clofibrate, fenofibrate, gemfibrozil, DEHP, bisphenol-A, estradiol and octylphenol) into groups that mirror their receptor interactions. The gene expression changes that drive these groupings include the fatty acid oxidation genes. DrugMatrix can further distinguish unique qualities of a single drug within a particular class. Gemfibrozil is contraindicated in hypertriglyceridemic patients due to an unusual elevation of LDL levels. Our studies show that it uniquely elevates fatty acid and cholesterol biosynthesis gene expression. These observations provide a molecular distinction that correlates with the clinical distinction of gemfibrozil from the other compounds and suggest a mechanism behind this distinction. These findings illustrate the value of an early, detailed understanding about adverse events and unappreciated benefits, speeding lead prioritization and selection

Key words. Chemogenomics, predictive toxicology, toxicogenomics, gene expression, molecular pharmacology

Introduction

The drug discovery industry has registered fewer and fewer new drugs each year of the past 20 years. Investigation of the elements that have reduced drug discovery success indicate that ~51% of drug development failures can be attributed to lack of efficacy, 30%, toxicology in animal models, 11%, and adverse events in humans, 10 % (198 NCE's taken into man, S. Albertini, SMI Pharmaceutical Conference, April 3 & 4, 2000, London). All of these issues arise from a poor understanding of the action of many drugs, especially novel leads and drug classes. Many investigations are focused on developing a better understanding of how drugs produce efficacy and cause adverse events.

Combining individual types of data, such as gene expression data, pharmacological assays, chemical SAR analysis, and physical-chemical properties of compounds in a unified contextual database allows a complete and integrated meta-analysis, increasing the value over analysis of each information domain independently. Creation of new high quality, complete and homogenous pharmacology and expression data also increases the value over reliance on literature findings. The integrated, homogenous view of a candidate's likely biological effects will translate into better candidate prioritization and better clinical success.

Iconix Pharmaceuticals is developing DrugMatrix™, an integrated database relating these data dimensions: molecular pharmacology profiles, gene expression profiles, chemical structure and SAR, toxicology, pathology, and pharmacology. DrugMatrix focuses on a diverse set of 2000 compounds, including all marketed drugs, late stage failures, standard reference toxicants, and well characterized biochemical ligands. Each compound is being analyzed by gene expression analysis across ~10,000 rat genes, using multiple animals and multiple times and doses. In addition, all compounds are being analyzed in a panel of ~130 pharmacological and drug metabolism assays. Finally, expert curation of chemical, toxicological, pharmacological and clinical data provides context and enhances the interpretation of experimental data. The combination of these new data sets creates a unique interpretive environment.

The final critical element of DrugMatrix is a specialized software interface that brings order and accessibility to the data. The software provides the ability to directly query the database, 1/2 billion data points, and ask questions such as: What compounds have the most similar expression patterns and/or most similar molecular pharmacology activities as my candidate drug? The scientist can then seamlessly move between domains within DrugMatrix to understand the compound in the context of gene expression, pharmacological profile, or chemical data. Without such software, a large reference database, like DrugMatrix, would only be accessible to the bioinfomatics specialist. Using DrugMatrix at the front line of the drug discovery scientists can derive new information and make new deductions and thus understand better and more rapidly optimize their lead compounds.

In this paper we illustrate the use of DrugMatrix to discriminate the mode of action of nuclear receptor agonists (PPARalpha, estrogen receptor and PXR receptor agonists) and correlate some of that data to a well-known adverse drug event.

Materials and Methods

Animals: Sprague-Dawley Crl:CD(SD) BR strain (VAF plus) rats aged 4-6 weeks were fed a standard rodent diet and allowed tap water *ad libitum*. Animal procedures were carried out at Sequani Ltd. (Ledbury, Herefordshire, England).

Compound dosing: All compounds were administered to groups of two male and two female rats for each dose and time. Estradiol benzoate, bisphenol A

(BPA) and octylphenol (OP) were administered sub-cutaneously in arrachis oil; clofibrate, fenofibrate, gemfibrozil and bis (2-ethyl-hexyl)phthalate (DEHP) were administered by oral gavage in 1% NaCMC. The doses chosen were the maximum tolerated dose (MTD), 70% MTD, 50% MTD and 10% MTD. The MTDs were determined from the literature or extensive internal experience at Sequani. The MTD's used were: estradiol benzoate, 2 mg/kg; BPA, 150mg/kg; OP, 450 mg/kg; clofibrate, 250 mg/kg; fenofibrate, 1000 mg/kg; gemfibrozil, 300 mg/kg; and DEHP, 1000 mg/kg. After dosing tissues were harvested at 3h, 24h or 72h. For the 3h and 24h time point, animals were dosed at time 0h and euthanized after 3 or 24 h. For the 72h group, animals were dosed at 0h, and at 24h and 48h. Tissues were collected and frozen on dry ice prior to storage at $-80^{o}C$.

RNA Isolation and Probe preparation: Homogenization of liver tissue, mRNA extraction, and probe labeling were performed as described (Yue et al., 2001). Each sample was hybridized to duplicate Rat Toxicology LifeArrays (Incyte Genomics, Palo Alto, CA) as described (DeRisi et al., 1997). The control mRNA was derived from a pool of livers from age- and strain-matched untreated animals (40 male and 40 female).

Data Handling and Analysis: All 680 microarrays were analyzed simultaneously after mean total signal intensity normalization across both channels as provided in GEM Tools®. Zeros replaced missing values.

Gene-by-Gene analysis: Ratios (ln, $\log_{10}$ or $\log_2$) for each clone were averaged across the same dose, time point, and sex. Uncentered-complete hierarchical clustering using the algorithm provided by Cluster and the visualization tool TreeView (Eisen et al.., 1998; http://rana.lbl.gov/EisenSoftware.htm). Genes that were differentially expressed by a drug or multiple drugs were identified visually and confirmed by determining the ratios that were significantly different between drug-treated animals and vehicle only-treated control animals using a Student's two-tailed t-test.

Distinctiveness analysis: Data were averaged over duplicate arrays and cropped to that set of expression ratios that were differentially expressed 2-fold or more at least once across all the conditions. The resulting 8,291 clones were analyzed further using the Distinction Calculation (Golub et al.., 1999) as implemented in Spotfire™ DecisonSite 6.3.

Principal Component Analysis (PCA): The 200 most variable genes as determined by the standard deviation of the ratios of a clone across all 680 experiments were selected as variables for PCA using Spotfire™ DecisionSite™6.3. Sorting according to their Eigen value for each PCA dimension identified important genes.

Bioassays: Molecular pharmacology assays were performed on all of the compounds in 130 different assays selected from the MDS-Panlabs catalog. The panel of assays was chosen to include important sites of drug action and drug toxicity. Those compounds that exhibit a fractional inhibition of ≥50% at 30 mM in a preliminary duplicate test were further studied using an eight-point triplicate concentration titration at 1/2 –log intervals from 30 mM, to determine an IC_{50} value.

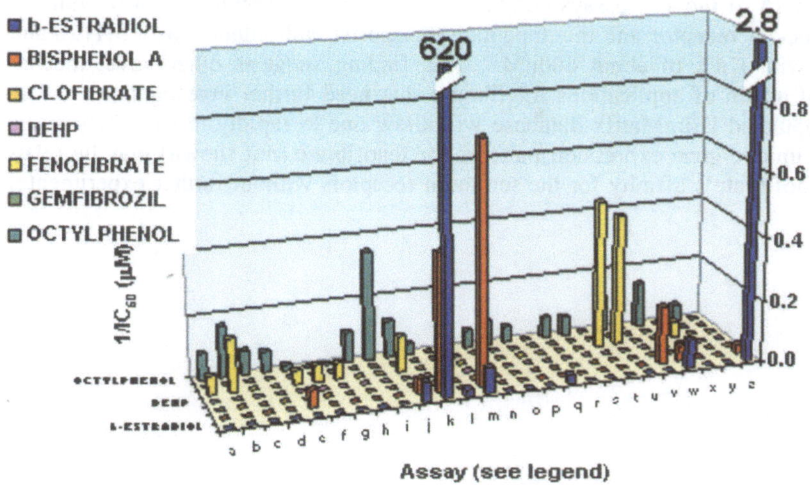

Fig. 1: Assay of biochemical activities of seven nuclear receptor agonists– estradiol, BPA, clofibrate, DEHP, fenofibrate, gemfibrozil and OP (reading front to back, z-ordinate) in 26 biochemical assays. These twenty-six bioassays were selected for display (from 123 assays) if any activity was detected in that assay for any of the 7 drugs. The x-ordinate represents biochemical assays as follows: a, acetylcholinesterase; b, adenosine A3; c, adrenergic alpha2Bd, adrenergic alpha2C; e, norepinephrine transporter; f, Ca-channel-L, dihydropyridine; g, Ca-channel-L, phenylalkylamine; h, cyclooxygenase COX-1; i, cyclooxygenase COX-2; j, dopamine transporter; k, estrogen ERalpha; l, glucocorticoid receptor; m, lipoxygenase 15-LO; n, muscarinic M1; o, muscarinic M2; p, muscarinic M3; q, progesterone receptor; r, S/T kinase, p38alpha; s, Y Kinase, EGF receptor; t, serotonin 5HT2A; u, serotonin 5HT2C; v, serotonin 5HT6; w, serotonin transporter; x, sodium channel, Site 2; y, tachykinin NK2; z, testosterone receptor. Each compound measurement that showed <50% inhibition in the preliminary test is approximated as zero.

Results and Discussion

Figure 1 illustrates a snapshot of the molecular pharmacology data from Drug Matrix for the seven compounds studied. These assays detect estronergic agents interacting potently with estrogen receptor-alpha, beta-estradiol (1.6 nM) and BPA (2.3 μM) beta-estradiol also interacts with the testosterone receptor (360 nM), and BPA interacts with 15-lipoxygenase (1.3 μM). These observations overlap known activities for ERalpha and testosterone receptor, while the 15-LO activity of BPA is novel.

Gemfibrozil, clofibrate and DEHP showed no activity in the 123 assays conducted to date. In contrast, octylphenol (OP) is quite promiscuous and interacts with 18 of the 123 assays conducted. Fenofibrate interacts weakly with the adenosine-A3 receptor and the dopamine transporter and potently with 5HT2a and 5HT2c with Kd's of about 600nM. This finding suggests other novel mechanisms of action of applications for fibrates that need further investigation. The fully populated DrugMatrix database will allow one to rapidly test the hypothesis that the unique gene expression patterns for fenofibrate (not shown) may be related to fenofibrate's affinity for the serotonin receptors without further experimental work.

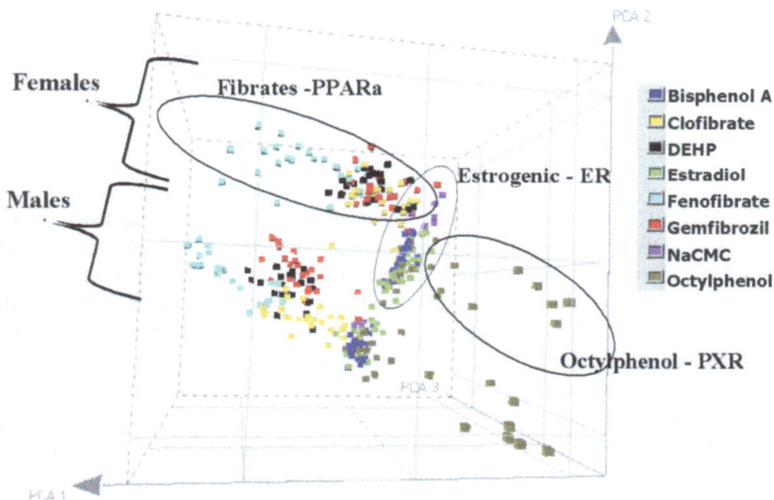

Fig. 2: Principal components analysis was performed on 680 individual microarrays using 200 gene variables. The first three principle components were visualized in the figure (labeled PCA1, PCA2 and PCA3). The result was colored according to the drug administered in each experiment. For clarity, only 340 experiments are shown corresponding to one of the duplicate arrays analyzed for each rat. The major groups visible in the graph are circled. Principal component analysis (PCA) was applied to gene expression data, which was composed of 680 arrays. Cumulatively the first three components account for ~60% of the total variance in the data set; the 1st, 2nd and 3rd principle component account for 27, 25 and 7%, respectively.

Figure 2 shows that the expression patterns cluster, using principal component analysis (PCA), into several distinct groups. Fibrates and other peroxisomal proliferator compounds such as DEHP cluster in one group while estradiol benzoate and BPA, both pure estrogen receptor (ER) agonists, and the vehicle controls group together. OP, a weak ER agonist that also has activity on the PXR, separates from the other compounds in a unique position. Each of these clusters are further split by gender.

The first principal component is dominated by the effect of peroxisomal proliferator PPARalpha agonists (the fibrates and DEHP), and is associated primarily with fatty acid beta oxidation gene expression. The effect of gender on the expression of certain genes, particularly 4-12 X-chromosome specific transcripts and certain sex steroid metabolism genes dominate the second principle component. The third principle component was dominated by the effect of OP (PXR/ER mixed agonist), and is associated with extracellular and blood protein genes that might be indicative of stress responses. The ER selective agonists (estradiol benzoate and BPA), and the vehicles are unresolved.

The separation of the PPARalpha agonists in one component, estradiol and BPA in another, and OP in a third correlates with the activities of these compounds on several receptors expressed in liver. DEHP and the fibrates potently stimulate PPARalpha, and DEHP, and fibrate liver toxicity requires PPARalpha (Corton et al., 2000; Ward et al., 1998; Kliewe et al., 1999), activities that correlate with the clustering of the PPARalpha agonists in the PCA (Figure 2). Estradiol stimulates the estrogen receptor with an ED_{50} near 10^{-11}M (Masuyama, 2000), while BPA, DEHP and nonylphenol (a very close relative of OP) stimulate the ER with EC_{50}'s of about 1 _M. DEHP and nonylphenol stimulate PXR receptors with an EC_{50} of about 0.5 _M, while estradiol and BPA are completely inactive on PXR (Masuyama 2000). ER active compounds (estradiol and BPA) clustered with the vehicle controls, perhaps because liver shows weak estrogenic response. Because DEHP, which is active on the PXR, did not induce the same genes as OP, OP's distinction may arise from activity on one or multiple other receptors. The potential activity of OP on other receptors is supported by its promiscuity in the molecular pharmacology assays (Figure 1) and reports in the literature (Masuyama, 2000).

To better understand which genes drive the discrimination of PPARalpha agonists (PP) from the ER and ER/PXR drugs (Non), the data were also analyzed using the discrimination metric developed by Golub *et al.* (Golub et al., 1999), Figure 3A. These calculations identified a number of genes that uniquely discriminate the PP set from the Non set. Of the top 100 most discriminating genes for PP, 35 were readily identified as in fatty acid beta oxidation (FABO) pathway and 25 were novel genes. We speculate that some of these 25 novel genes may be unappreciated members of the FABO pathway.

PCA and distinction calculations identify a very overlapping set of genes. Fourteen of the top 15 genes (93%) identified by PCA overlap with the top 100 most distinctive genes. The discrimination of PPARalpha agonists from the other drugs by two methods provides a cross validation of the results, suggesting that the FABO pathway is a defining effect of the PPAR-agonist drugs.

Are there unique gene expression patterns within a drug family? Figures 3B and 3C show examples of two genes looking across several different PP drugs, and demonstrates that there are genes that exhibit compound-specific profiles.

Many genes that were identified as PPAR agonist genes are upregulated across all PP compounds and across most time points and doses. For example, the peroxisomal bifunctional enzyme is significantly and rapidly elevated up to 10-fold over that of untreated by all PP compounds (DEHP not shown) (Figure 3B). In contrast, other genes such as fatty acid synthase are uniquely elevated by gemfibrozil only (Figure 3C). Detailed analysis using PCA, distinction calculation, and hierarchical clustering identified 13 genes specifically induced by gemfibrozil, but not other PPAR agonists. Eight of 13 genes are involved in cholesterol and fatty acid biosynthesis. The distinction of gemfibrozil from other fibrates in stimulating fatty acid anabolism genes correlates with a noted clinical contraindication. Fibrates are used to treat hyperlipoproteinemias, mainly by elevating the rate of fat oxidation in the liver (Kersten et al., 2000; Ward et al., 1998), a mechanism supported by the up regulation of the FABO pathway genes shown here. In many patients, especially hypertriglyceridemic patients, gemfibrozil, but not other fibrates, induces an increase in LDL levels (Mahley and Bersot 2001). Elevated fatty acid production raises VLDL and ILDL levels and subsequently LDL (Mahley and Bersot 2001). The observation that gemfibrozil increases fatty acid/cholesterol biosynthesis gene expression may provide a molecular explanation of the paradoxical clinical effect of this fibrate.

Conclusions

These studies show how the combination of molecular pharmacology and gene expression analysis clearly separates several different nuclear receptor agonists into drug families. These separations place the PPARalpha agonist in one family and estrogenic agents in another family. OP is distinct from these two families and interacts promiscuously with various receptors and enzymes. Further subdivisions of each drug family is possible by extending the distinction calculation analysis of the expression data to identify compound-specific, as well as class-specific, gene expression profiles. Our studies provide a molecular distinction that proposes a mechanism behind the clinical distinction of gemfibrozil from the other fibrates. The DrugMatrix approach can significantly benefit the front line of drug discovery by providing early access to data about possible adverse events and unappreciated benefits of new candidate drugs, hastening lead selection and improving lead prioritization, and increasing the efficiency with which new drugs reach the market.

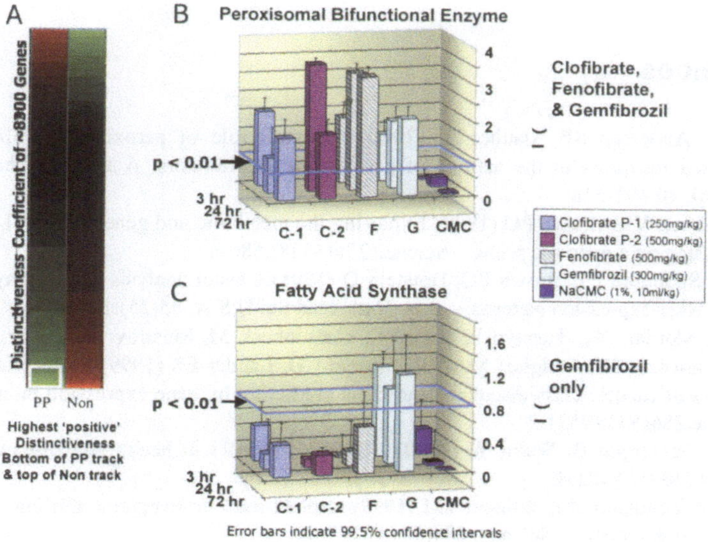

Fig. 3: Identification of genes that separate PPAR-alpha agonists from ER and PXR agonists. A, The four drugs that induce peroxisome proliferation and are PPARalpha agonists (labeled as PP) were included in one group and compared to a second group containing beta-estradiol, BPA, and OP and the vehicle-treated samples (labeled as Non). These groups were analyzed using the Distinction Calculation tool in Spotfire™ DecisionSite 6.3 on all the genes. The highest positive Distinction Coefficient Non group was used to arrange the ~8200 genes. The white-boxed group of genes, left track bottom, marked PP, has the highest distinctiveness for PPAR-alpha agonists. B & C, Examination of individual genes within the PPAR-alpha discriminating gene group, uncovers genes with several patterns of expression for individual drugs within the PPARalpha drug family. B, Example of a gene that responds to all family members, peroxisomal bifunctional enzyme. C, Example of a gene, fatty acid synthase, which responds to only a single drug family member, gemfibrozil. The plane, labeled $p<0.01$, represents the mean expression level necessary to achieve a t-test figure of merit of <0.01. In panel B all of the drug-treated measurements are significant at $p<0.001$ except Gemfibrozil-24 hr, $p<0.05$, and Fenofibrate-3 hr, ns. In panel C only the Gemfibrozil measurements are significant, 72 hr and 3 hr., $p<0.005$, and 24 hr, $p<0.01$. The clofibrate treatment was preformed twice at two different doses, 250 mg/kg, corn oil (P1) and 500 mg/kg, 1% NaCMC (P2).

References

Corton JC, Anderson SP, Stauber A (2000) Central role of peroxisome proliferator-activated receptors in the actions of peroxisome proliferators. Annu Rev Pharmacol Toxicol. 40:491-518

DeRisi JL, Iyer, VR, Brown PO (1997) Exploring the metabolic and genetic control of gene expression on a genomic scale. Science 278(5338):680-6

Eisen MB, Spellman PT, Brown PO, Botstein D (1998) Cluster analysis and display of genome-wide expression patterns. Proc Natl Acad Sci U S A. 95(25):14863-8

Golub TR, Slonim DK, Tamayo P, Huard C, Gaasenbeek M, Mesirov JP, Coller H, Loh ML, Downing JR, Caligiuri MA, Bloomfield CD, Lander ES (1999) Molecular classification of cancer: class discovery and class prediction by gene expression monitoring. Science 286(5439):531-7

Kersten S, Desvergne B, Wahli W (2000) Roles of PPARs in health and disease. Nature. 405(6785):421-4

Kliewer SA, Lehmann JM, Willson TM (1999) Orphan nuclear receptors: shifting endocrinology into reverse. Science. 284(5415):757-60

Mahley, RW, and Bersot, TP (2001) In: Hardman, JG, Limbird, LE, (Eds) Goodman's and Gilmans's: The Pharmacological Basis of Therapeutics, 10th ed., McGraw Hill pp 971-1002

Masuyama H, Hiramatsu Y, Kunitomi M, Kudo T, MacDonald PN (2000) Endocrine disrupting chemicals, phthalic acid and nonylphenol, activate Pregnane X receptor-mediated transcription. Mol Endocrinol. 14(3):421-8

Ward JM, Peters JM, Perella CM, Gonzalez FJ (1998) Receptor and nonreceptor-mediated organ-specific toxicity of di(2-ethylhexyl)phthalate (DEHP) in peroxisome proliferator-activated receptor alpha-null mice. Toxicol Pathol. 26(2):240-6

Yue, H, Eastman, PS, Wang, BB, Minor, J, Doctolero, MH, Nuttall, RL, Stack, R, Becker, JW, Montgomery, JR, Vainer,M and Johnston, R (2001) An evaluation of the performance of cDNA microarrays for detecting changes in global mRNA expression. Nucleic Acids Research 29(8): E41-1

Reverse toxicology as a future predictive toxicology

Jun Kanno

Cellular and Molecular Toxicology Division, Biological Safety Research Center, National Institute of Health Sciences

Summary. The whole genome sequence of humans and rodents, and technical capability of monitoring the whole genome expression are now in our hands. In theory, "whole genome profiling" is a way to reveal the molecular events taking place inside our body or in experimental animals. The major characteristic of the whole genome profiling is that it does not require overt phenotyping *per se*. In other words, it is possible to build up a gene expression database regardless of the presence or absence of the classical toxicological phenotypes. This phenotype-independent approach can be considered "reverse toxicology" in contrast to the traditional toxicology or "forward toxicology", a phenotype-dependent approach currently performed as a standard procedure. The future of predictive toxicology would be based on a well-balanced informatics system composed of both the reverse and forward toxicology. This approach is intended to clarify, 1) the molecular based mechanisms of species/strain differences that will allow more accurate extrapolation of animal model data to humans and 2) a toxicogenomics-based risk assessment strategy that doesn't overlook "unexpected toxicity". In conclusion, in the future, toxicology will be fully developed as a mechanism-based science through the process of disclosure of the "black box" of organisms by coordinated implementation of reverse toxicology and forward toxicology.

Key words. Whole genome sequencing, DNA microarray, comprehensive gene expression monitoring, reverse toxicology

Toxicology has been a science based on phenotypes induced by exogenous agents in an organism. Toxic symptoms are categorized and the causative agents are classified in a symptom-based fashion. When human data are not available, rodent models as surrogates have been used to collect phenotype data for human risk assessment. This approach has served us well for a long time with a series of well-considered strategies including a concept of "safety factors" or "uncertainty factors". This empirical and practical approach has been based on the scientific knowledge available (Fig 1). However, because the strategy has been dependent on measurable phenotypes, it has not always been perfect. One well-known

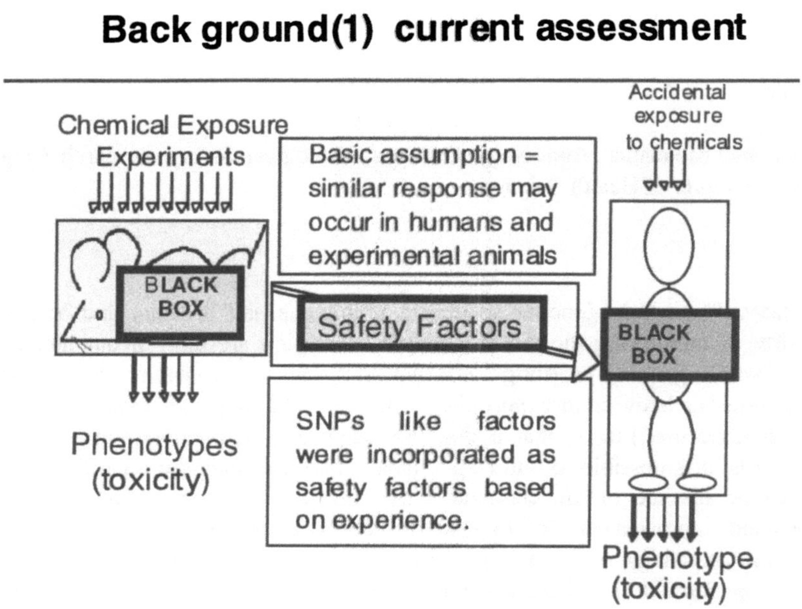

Fig. 1. The use of "Safety Factors" in traditional toxicology is based on available scientific knowledge of toxic phenotypes. This scheme has been robust as long as the black boxes are similar in quality in response to the xenobiotic challenges.

example is the thalidomide story, which induced no teratogenic responses in rat and mouse models, but did induce teratogenic effects in humans and rabbits. In order to improve the extrapolating process from rodents to humans, it is clear that a mechanism-based approach must be reinforced. Recent studies on thalidomide-regulated genes strongly suggest a possibility that teratogenic effects not observed in rodents can be monitored as angiogenesis-regulating properties even in the rodent models themselves (1,2). Once the responsible gene(s) or those closely related to a particular function/phenotype are identified in both species, extrapolation of the biological effects can be achieved with increased accuracy.

To identify responsible genes in mice, gene knockout (KO) technology can be applied, and in humans, single nucleotide polymorphism (SNPs) data can be utilized (Fig. 2). The limitation, however, is that these approaches are still phenotype-dependent. In other words, if there is no observable phenotype in a KO mouse, it is usually a "dead end" for further analysis on the function of the targeted gene. The same situation is usually true when the SNPs are not well linked to a distinct symptom. Utilization of gene expression profiling by DNA microarray technique as a screening strategy is another way to identify mechanistically relevant genes. This approach is also dependent on phenotype and thus has the same limitation mentioned above.

To overcome the "dead end" situation in the case of "no observable phenotype", our proposal is to introduce a "reverse approach" paradigm. The approach is to construct a bioinformatics database dependent upon a gene expression profile generated even in the absence of a xenobiotics-induced phenotype. This approach is now possible because of two factors: 1) the

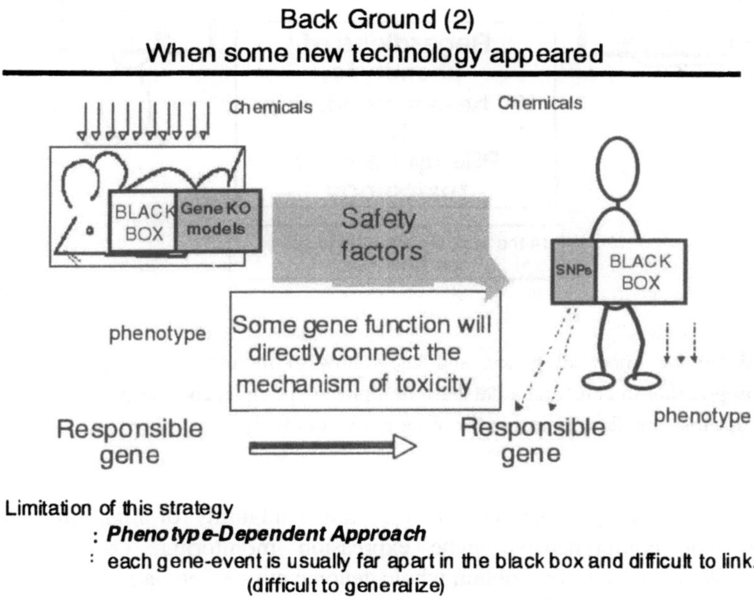

Fig. 2. Finding the responsible gene for a particular phenotype is possible by using gene knockout mice or SNPs (single nucleotide polymorphism) data in humans, as well as an attempt to screen out related genes by DNA microarray techniques. However, all these approaches are still dependent on phenotypes. If there is no phenotype in relation to treatment or genotype, the search for responsible gene usually becomes very inefficient.

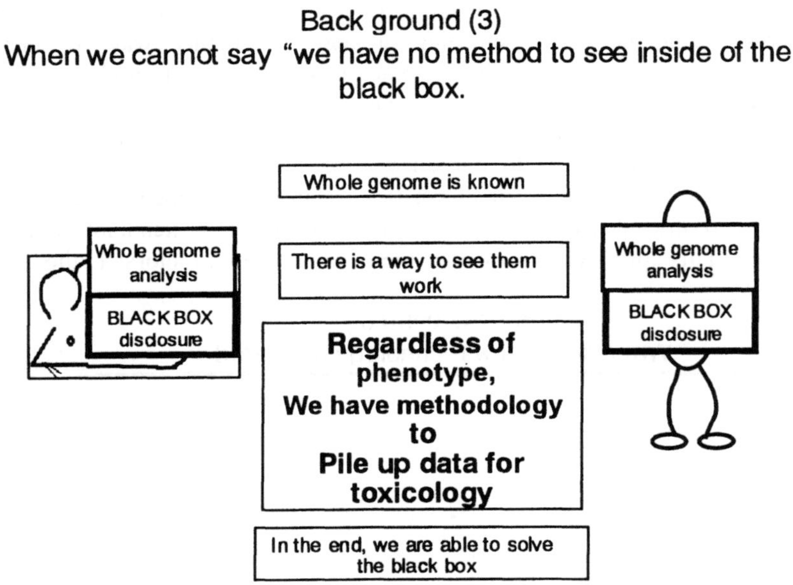

Fig. 3. When we know all genes, and know how to monitor their expression, it is theoretically possible to construct a database in a phenotype-independent approach that will eventually disclose the network inside the black box of our body.

availability of whole genome data, and 2) the availability of high throughput technology for comprehensive gene expression monitoring, i.e. the DNA microarray. We propose to first obtain whole genome expression data, and then to analyze what phenotype, if any, could be linked to the treatment-related gene expression profiles (Figure 3). This approach could be most effectively applied to the low dose issue in endocrine disruptor studies. In the endocrine system, exogenous perturbation is effectively cancelled out by homeostatic feedback mechanisms. Even though exogenous stimuli may fail to induce overt phenotype, such stimuli should clearly induce gene expression involved in a homeostatic response (Fig 4).

The DNA microarray technique is applicable to both a forward and reverse approach. In the forward approach, genes that are significantly altered in expression are linked to an overt phenotype. Databases built on such selected expression data are currently available and growing fast. In contrast, in the reverse approach, expression data on all genes, including those not linked to an observable phenotype, are equally important and incorporated into the database. To assure the quality of the database, strict normalization of gene expression levels is essential. In other words, unchanged gene expression should be monitored correctly as

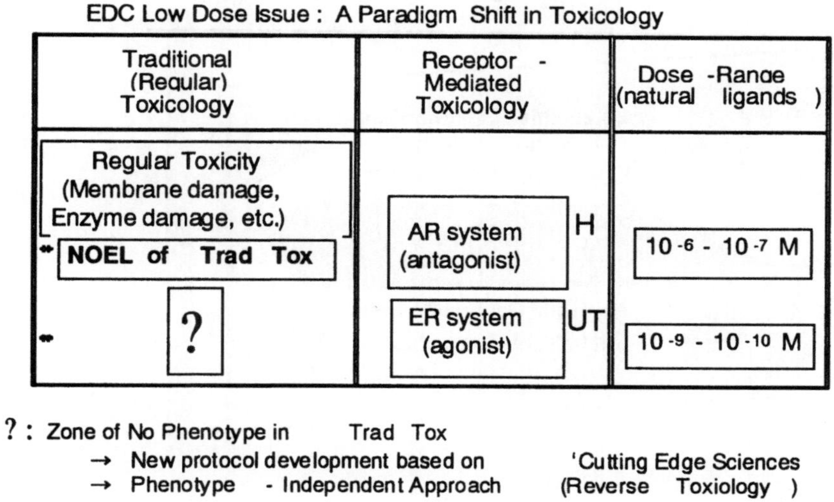

Fig. 4. The low dose issue of endocrine disrupting chemical research is the problem of gap between monitorable does-range in traditional toxicology and receptor mediated toxicology. Estrogenic responses monitorable, at least, by uterotrophic assay (UT) and some of the androgenic responses monitorable by Hershberger assay (H) are below the level of no observed effect level (NOEL) of traditional toxicology studies. The range indicated by "?" is the zone of no-phenotype in traditional toxicology. There seems to be two ways to handle this "?" zone, 1) develop new test protocols by cutting edge sciences, 2) apply reverse toxicology as phenotype-independent approach.

"unchanged". Reverse toxicogenomics requires continuity and compatibility of the data for its compilation. It includes a strategy to prepare a standard sample set to normalize various gene-expression monitoring platforms including SAGE, quantitative PCR, DNA microarray systems of different platforms, including different microarray versions in a particular system, different reading machines, and various microarray stocks, buffer stocks, etc.

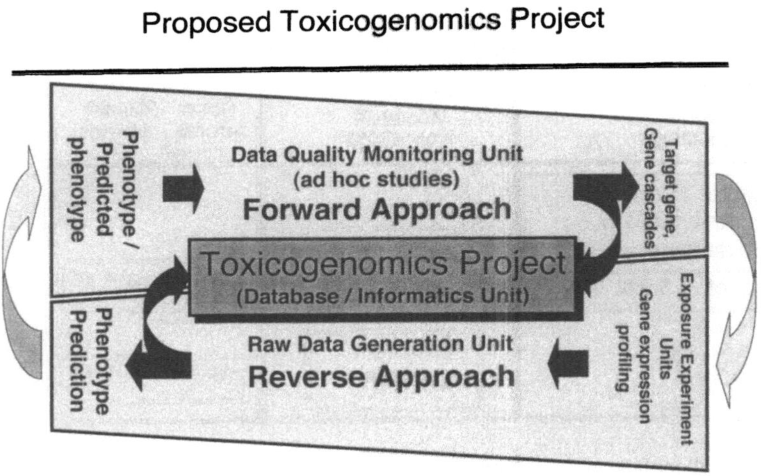

Fig. 5. Proposed Toxicogenomics Project is characterized by its components, the Reverse Approach and the Forward Approach. The database/informatics are built on data generated from both approaches.

The reverse approach will generate an enormous amount of raw data. A set of gene expression data that is linked to known phenotypes is essential for the startup of the toxicogenomics database. Then, gene clusters of unknown function will be investigated by a forward approach to connect the missing link to known gene clusters. This exploratory approach to functionally characterize unknown gene cluster demands ad hoc experiments including gene-knockout animal studies designed wisely by a multidisciplinary team of molecular biologists, chemists, toxicologists, pathologists, etc. (Fig 5.) The final goal is to fully develop a mechanism-based predictive toxicology through the process of disclosure of the "black box" of organisms by coordinated implementation of reverse toxicology and forward toxicology.

Index

D

M

N

Q

R

S

T

U

V

W

X

Y

MIX
Papier aus verantwortungsvollen Quellen
Paper from responsible sources
FSC® C105338

If you have any concerns about our products,
you can contact us on
ProductSafety@springernature.com

In case Publisher is established outside the EU,
the EU authorized representative is:
Springer Nature Customer Service Center GmbH
Europaplatz 3, 69115 Heidelberg, Germany

Printed by Libri Plureos GmbH
in Hamburg, Germany